Bio-Inspired Wettability Surfaces

Bio-Inspired Wettability Surfaces

Developments in Micro- and Nanostructures

Yongmei Zheng

Pan Stanford Publishing

Published by

Pan Stanford Publishing Pte. Ltd.
Penthouse Level, Suntec Tower 3
8 Temasek Boulevard
Singapore 038988

Email: editorial@panstanford.com
Web: www.panstanford.com

British Library Cataloguing-in-Publication Data
A catalogue record for this book is available from the British Library.

Bio-Inspired Wettability Surfaces: Developments in Micro- and Nanostructures

ISBN 978-981-4463-60-7 (Hardcover)
ISBN 978-981-4463-61-4 (eBook)

Printed in the USA

Contents

Preface

There has been an exciting confluence of research areas, such as physics, chemistry, biology and materials science, in recent years. A kernel consists in organic materials with high/low surface energy, has various-style micro- and nanostructures, such as regular/irregular, ordered/disordered, rough/smooth that can be endlessly arranged and combined, and is greatly provided with adaptability to perfectly display the biological functions that developed during the thousands of years evolution in nature. Interestingly, the biological surfaces develop the micro-/nanostructure with gradient features to smartly achieve the wetting controls. For example, the ultra-hydrophobic water repellency on lotus leaf; directional water collection on wetted spider silk; direction adhesion of super-hydrophobic butterfly wing, and fog-collecting hydrophobic/hydrophilic pattern on beetle back. Biological surfaces provide endless inspiration for the design and fabrication of functional interface materials with unique wettability, generating promising applications such as micro-fluidic devices, functional textiles, corrosion resistance, liquid transportation, and anti-fogging and water-collecting devices.

Researches on wettability, water repellency, bioinspiration, and biomimetics have shown an increasing trend, and has includeed realms of materials science, chemistry, physics, engineering, polymer science, science technology, metal engineering, mechanics, energy fuels, geology, and so on. On the global front, China has published the largest number of results in recent years of the researches conducted and products developed.

The book introduces the recent researches on wettability of biological and bio-inspired surfaces through its four chapters. The first chapter focuses on the wettability features and effect in a lotus leaf. It discusses the basic concept of wettability and suggests methods for fabrication bio-inspired surfaces and achieving functional surfaces. In the second chapter, wettability features and effect in a butterfly wing are described. It specifically stresses on the anisotropic wettability resulting from anisotropically structured surfaces. Anisotropy ranges from one dimension to two dimensions

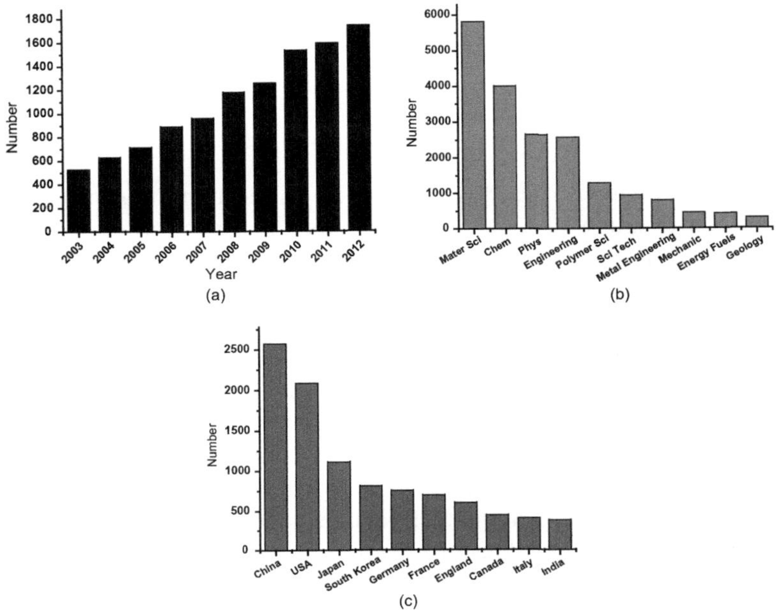

Number of researches on wettability, water repellency, and bio-inspired and biomimetic materials (a) in recent years, (b) in different areas, and (c) in various countries.

on a surface for droplet movement. Many fabricating methods of bio-inspired surfaces can be developed further to obtain fluid-controlling functional surfaces. In the third chapter, the wettability features and effect of spider silk are introduced and novel mechanisms of water collection resulting from gradient multi-structures have been described. Some methods to fabricate bio-inspired materials that were developed by our team have also been introduced. Finally, the fourth chapter presents the wettability features and effect of beetle back for fog collection and discusses about the methods for fabricating hydrophobic/hydrophilic patterns.

The book offers the mechanisms of smart wetting control, such as water collection and repellency, on biological micro-/nanostructure gradient interfaces that were developed by the us in recent years. It also suggests ideas to mimic biological features to realize bio-inspired functional surfaces with unique wettability. The book will help researchers to develop innovatory designs from novel materials for future scientific works.

The book is a result of the collective and creative research efforts of our team. I would like to thank Drs. Qunfeng Cheng, Yongping Hou, and Yuan Chen for their efforts and support in helping me write Chapters 1, 2, and 3 and 4, respectively. I especially thank Stanford Chong for offering me the opportunity to publish my favorite book.

I would also like to acknowledge the support given by the National Key Basic Research Program of China (2013CB933001), the National Natural Science Foundation of China (21234001, 21473007), and the Doctoral Fund of Ministry of Education of China (20121102110035) to bring out this book.

Yongwei Zhang

Summer 2015

Chapter 1

Lotus Leaf Effect: Micro- and Nanostructures

1.1 Janus Feature of Lotus Leaf

Inspired by the ancient Roman god Janus with two opposing faces, the Janus particles are firstly fabricated.[1] Then, the Janus fabrics,[2] Janus nanoparticles,[3] and nanosheets[4] are also demonstrated. In fact, this kind of Janus feature can be found in nature. Recently, we discovered that the lotus leaf shows the Janus feature: the surface of the back-side of a superhydrophobic lotus leaf can be easily completely wetted with water in the air and repels oil droplets in water, as shown in Fig. 1.1.[5] Three water droplets form spheres on the upper side of lotus leaf floating on the water surface, and three oil (n-hexane) droplets stay as perfect spheres under water on its back-side.

Figure 1.1 Digital photograph shows the Janus feature of lotus leaf flowing on the water surface: three water droplets form spheres on its upper side in the air, and three oil (n-hexane) droplets stay as perfect spheres under water on its back-side.

Bio-Inspired Wettability Surfaces: Developments in Micro- and Nanostructures
Yongmei Zheng

ISBN 978-981-4463-60-7 (Hardcover), 978-981-4463-61-4 (eBook)
www.panstanford.com

It is well known that the lotus leaf has attracted immense scientific interest due to its superior self-cleaning property (Fig. 1.2), exhibiting a high water contact angle (CA) above 160° and a small sliding angle (SA) ~2°.[6] Barthlott and Nienhuis first revealed that the large CA is attributed to the epicuticula wax and the micrometer-scale papillae structure of the lotus leaf surface.[7] Jiang et al. later found that there are the micro-/nanoscale hierarchical structures on the lotus leaf.[8] After considering the contribution of the nanostructures, the theoretical model calculation indicates the CA can be over 160°, which is well consistent with the observed experimental results. It is obvious that the upper side of lotus leaf with a typical micro-/nanoscale hierarchical papillae and the epicuticular wax makes raindrops easily roll across the upper side of lotus leaf carrying away dirt and debris. On the other hand, we found that there are many microfolds around the margin of lotus leaf in the form of ring-bands, introducing high-energy barrier against water to block the water underneath preventing overflow onto its upper surface.[9] Thus, the upper side of lotus leaf can keep itself clean from the dirt particles that are brought away when the water droplets roll off. This self-cleaning effect is usually called "lotus effect," which shows great importance in fundamental research and potential in industrial applications.

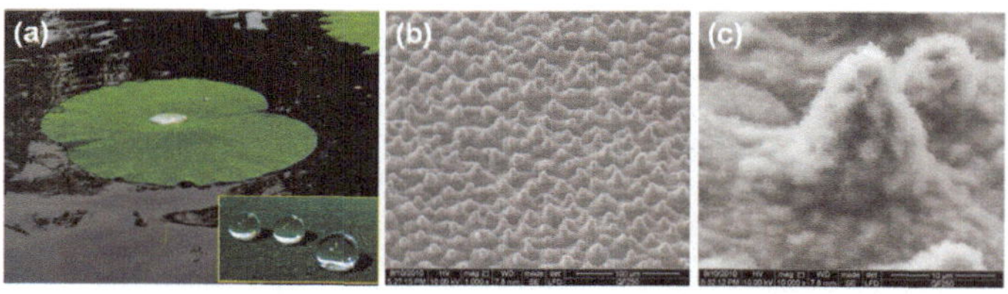

Figure 1.2 (a) Digital photograph of the floating lotus leaf on the water and water droplet on the superhydrophobic upper side of lotus leaf. (b) Low magnification SEM image of the surface morphology of the upper side of lotus leaf. (c) High-resolution SEM (HRSEM) image of a single papilla with branch-like nanostructures.

Inspired by micro-/nanoscale hierarchical structure of the upper side of lotus leaf, many methods are developed to construct the superhydrophobic self-cleaning surfaced, such as phase separation,[10,11] oxygen plasma treatment,[12] sol–gel foam,[13] electrospinning,[14] self-assembly technique,[15] and so on. Because there are many reviews that have reported these methods,[6,16–18] in this chapter, we will focus on the lower side of lotus leaf.

In fact, the lotus leaf comes out of the mud soiled and remains undefiled spite of general corruption, depends not only on the upper side but also the lower side of lotus leaf. Usually, the property of lower side of lotus leaf is often neglected. In this chapter, the surface morphology and wetting properties of lower side will be focused on. The environment scanning electron microscope (ESEM) images show that its lower side is made of many cells (Fig. 1.3a), and every cell consists of numerous tabular and slightly convex papillae with 30–50 μm in length and 10–30 μm in width (Fig. 1.3b). The single papilla is further tested by atomic force microscope (AFM). Every single papilla is covered with nanogrooves structure with a size of 200–500 nm and the height of single papilla is around 4 μm (Fig. 1.3c). Moreover, the epidermal glands of its lower side may secrete some hydrophilic compounds, such as the ferns.[19]

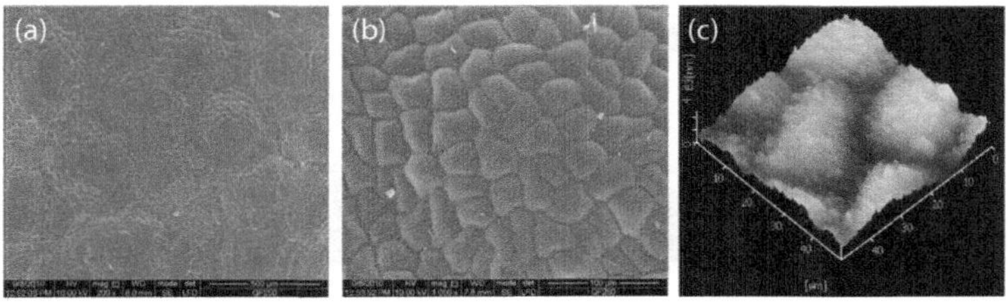

Figure 1.3 Environment SEM (ESEM) images (a & b) show its lower side is made of many cells, and every cell consists of numerous tabular and slightly convex papillae with 30–50 μm in length and 10–30 μm in width. (c) Atomic force microscope (AFM) image further shows the tabular papillae are covered with nanogrooves structure with a size of 200–500 nm and the height of single papilla is around 4 μm.

The lower side of lotus leaf can be completely wetted with water in the air due to absence of three-dimensional wax crystals on its lower side, similar to other species grown completely in water or partially floated on the water surface. However, under water the lower side of lotus leaf shows superoleophobic properties for apolar and amphiphilic oils, including 1,2-dichloroethane, *n*-hexane, octanol, sunflower oil, and rapeseed oil; and the stability of superoleophicity is very well after keeping it underwater for 24 h. All the oil contact angles (OCA) are over 150° as listed in Table 1.1.

Table 1.1. The oil contact angles (OCAs) of apolar and amphiphilic oils, such as 1,2-dichloroethane, *n*-hexane, octanol, sunflower oil, and rapeseed oil on lower side of lotus leaf is over than 150°, with keeping 24 h under water, showing the stability of superoleophobicity of its lower side is very well

Oil droplets	1, 2-dichloroethane	*n*-hexane	octanol	sunflower oil	rapeseed oil
In water (0 hour)					
	155.0 ± 1.5°	153.5± 1.5°	152.1± 1.5°	158.8 ± 4.6°	154.2 ± 1.2°
In water (24 hours)					
	154.0 ± 3.9°	152.2 ± 1.9°	150.9 ± 1.4°	153.2 ± 2.5°	151.9 ± 2.0°

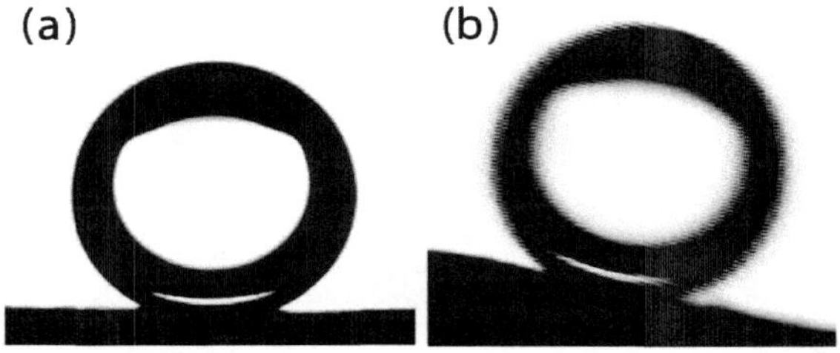

Figure 1.4 (a) Lower side of lotus leaf shows superoleophobicity with oil contact angle (OCA) of 155.0 ± 1.5° for 1, 2-dichloroethane. (b) When the tilt angle reaches about 12.1 ± 2.4°, the oil droplet easily rolls off from its lower side.

On the other hand, we also studied the sliding property of oil droplet on its lower side, take 1,2-dichloroethane as an example. The tilt angle with oil is about 12.1 ± 2.4°, with the OCA of 155.0 ± 1.5°, as shown in Fig. 1.4. The oil droplets can easily roll off on its lower side when the lotus leaf swings with external wind, and this means that its lower side is very difficult to be contaminated by the oil droplets under water and keeps itself clean in the mud.

1.1.1 Mechanism of Underwater Superoleophobicity

The wettability of solid surface is commonly evaluated by the Young's equation, which was originally applied in air. For superoleophobic surface in water, the modified equation can be inferred from Young's equation (1), as follows:

$$\cos\theta_3 = \frac{r_{l1-g}\cos\theta_1 - r_{l2-g}\cos\theta_2}{r_{l1-l2}} \qquad (1)$$

$$\cos\theta_3' = f\cos\theta_3 + f - 1 \qquad (2)$$

where θ_1, θ_2, and θ_3 are the contact angle of liquid 1 in air, liquid 2 in air, and liquid 1 in liquid 2, respectively. r_{l1-g}, r_{l2-g}, and r_{l1-l2} are the liquid 1/gas interface tension, the liquid 2/gas interface tension, and the liquid 1/liquid 2 interface tension, respectively, as shown in Fig. 1.5. Herein, the liquids 1 and 2 represent oil and water, respectively. On the other hand, the surface tension of water (r_{l2-g}) is much higher than that of oil (r_{l1-g}). Thus, for the hydrophilic surface, $\cos\theta_2$ is positive and the $\cos\theta_3$ can be inferred to be negative, which means hydrophilic surface in air should show oleophobic in water. If the rough structure is introduced on the hydrophilic surface, the superoleophobic surface in water can be obtained. The traditional Cassie model can be inferred as in equation (2). f is the area fraction of solid contacting with oil, and θ_3' is the contact angle of an oil droplet on a rough surface in water. And θ_3' becomes larger with the f decreasing, which means the water molecules will be much and easily trapped in the hierarchical micro-/nanoscale structure. The trapped water will prevent the penetration of oil droplets, resulting in superoleophobic and low-adhesive surfaces in water. In other words, the hydrophilic chemical compositions and hierarchical micro-/nanoscale structures are two key parameters to design superoleophobic surface in water.

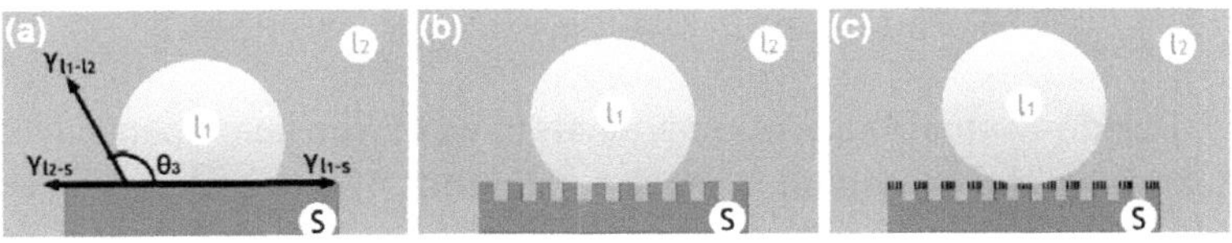

Figure 1.5 Schematic illustration of the effect of surface structure on the wetting behaviors of solid surface in oil/water/solid three-phase systems. (a) Smooth surface, (b) microscale structure surface, and (c) micro-/nanoscale hierarchical structure surface in liquid 2 phase.

1.1.2 Other Underwater Superoleophoic Surface

According to the aforementioned mechanism of superoleophobicity in water, the hydrophobic surface with hierarchical micro-/nanoscale structure would show superoleophobic property. Besides the lotus leaf, other water plants and animals will be briefly discussed in this chapter. For example, the water plant named "star grass," belongs to *Eichhornia* sp. As shown in Fig. 1.6, the leaf of "star grass" shows superoleophobicity with OCAs of 158.6 ± 1.4°, and 155.5 ± 1.6°, respectively, for its upper side and lower side in water. SEM images indicate its leaf surface consists of numerous tabular and convex papillae with 40–60 μm in length and 20–40 μm in width. Obviously, this typical hierarchical micro-/nanoscale structure contributes its superoleophobicity for oil droplets in water. The tilt angle of its leaf is less than 5°, resulting in the contaminant easily rolling off from its leaf surface and itself keeping clean in water.

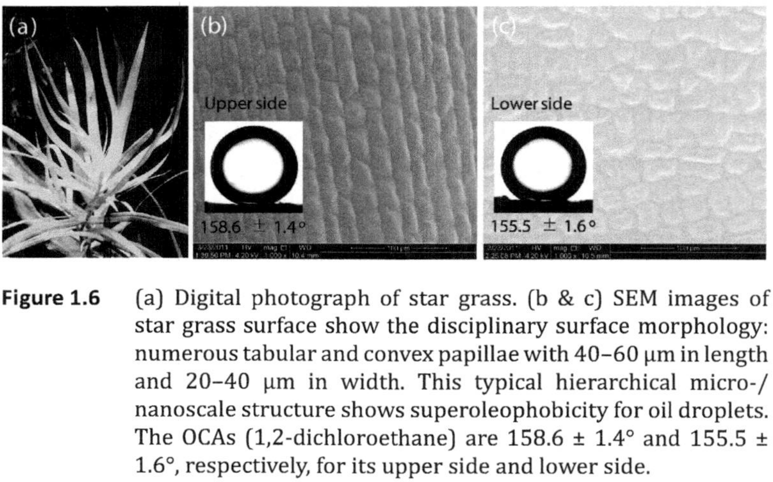

Figure 1.6 (a) Digital photograph of star grass. (b & c) SEM images of star grass surface show the disciplinary surface morphology: numerous tabular and convex papillae with 40–60 μm in length and 20–40 μm in width. This typical hierarchical micro-/nanoscale structure shows superoleophobicity for oil droplets. The OCAs (1,2-dichloroethane) are 158.6 ± 1.4° and 155.5 ± 1.6°, respectively, for its upper side and lower side.

Besides water plants, other animals in water can keep themselves clean from the contaminant, such as all kinds of fish scales,[20] loach skin,[21] and so on. Usually, the fish scale is typical organic–inorganic hybrid composites with calcium phosphate and protein, and coated with a thin layer of superhydrophilic mucus. SEM images (Fig. 1.7) show that the array-arranged sector-like fish scales with diameters of 4–5 mm is composed of oriented micropaillae with 100–300 μm in length and 30–40 μm in width, which arrange in the radial direction. Further observation shows the surface of micropapillae is nanoscale rough, which is confirmed with underwater AFM, as shown in Fig. 1.7f. The OCA of fish scale is about 156.4 ± 3.0° for oil droplets of 1,2-dichloroethane. Obviously, the underwater superoleophobic property is attributed to the hierarchical micro-/nanoscale structures on the fish scales.

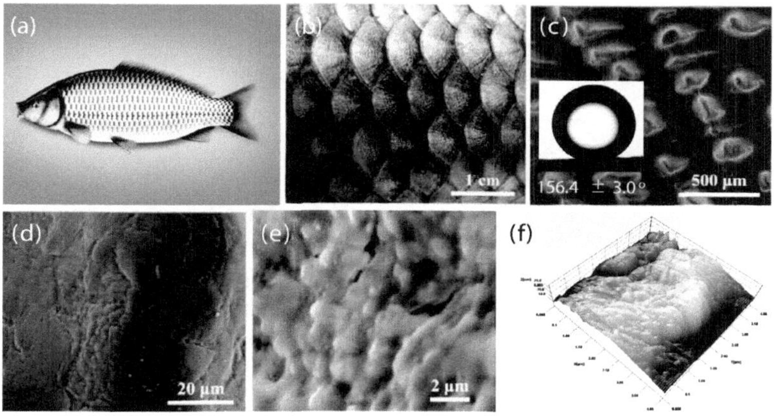

Figure 1.7 (a) The fish can keep their body clean in the mud water. (b & c) SEM observation revealed that the array-arranged sector-like fish scales with diameters of 4–5 mm is composed of oriented micropaillae with 100–300 μm in length and 30–40 μm in width, which arrange in the radial direction. The OCA of fish scale is about 156.4 ± 3.0° for oil droplets of 1,2-dichloroethane. (d & e) Further observation shows the surface of micropapillae is nanoscale rough. (f) Underwater AFM image further confirmed the nanoscale roughness of fish scale.

Compared with fish, loach prefers to live in the sludge. However, the loach can easily keep their body clean from contamination. The OCA of loach skin can reach as high as 160.9 ± 1.0° for oil droplets of 1,2-dichloroethane, as shown in Fig. 1.8. Surface structure observation revealed that its skin is composed of the typical hierarchical micro-/nanoscale structure. The length of the sector-like skin is 300–500 μm and the width is about 200 μm. Every sector skin is arranged of many ridges with 10–30 μm in length and 5–10 μm in width in the radial direction. Further AFM observation shows the surface of ridge is nanoscale rough, as shown in Fig. 1.8f.

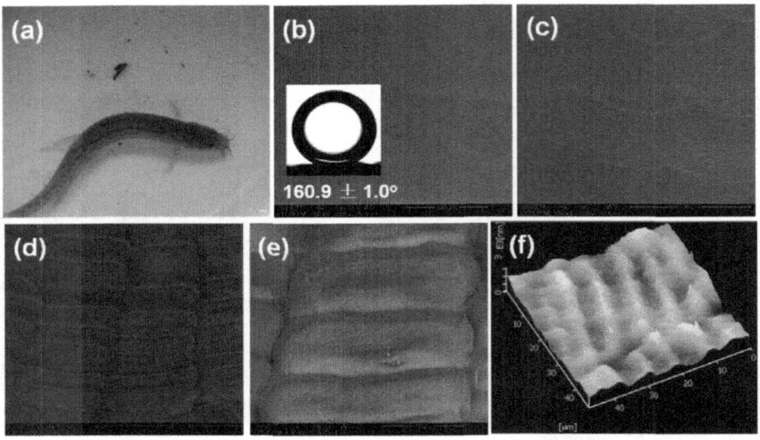

Figure 1.8 (a) The loach can easily keep their body clean in the sludge. (b–e) The OCA of loach skin can reach as high as 160.9 ± 1.0° for oil droplets of 1,2-dichloroethane. SEM images revealed that its skin is composed of the typical hierarchical micro-/nanoscale structure. The length of sector-like skin is 300–500 μm and the width is about 200 μm. Every sector skin is arranged of many ridges with 10–30 μm in length and 5–10 μm in width in the radial direction. (f) AFM observation shows the surface of ridge is nano-scale rough.

1.2 Fabrication Methods of Underwater Superoleophobic Surface

Constructing underwater superoleophobic surface is a new challenge for scientists. The traditional methods of fabricating superoleophobic surface in the air are dependent on modification of fluorinated materials, and the use of fluoride is harmful to the environment. On the other hand, the superoleophobic surface in the air usually shows superoleophilicity underwater.[22,23] Based on the aforementioned discovery in nature, several bio-inspired methods of constructing the fluoride-free underwater superoleopohic surface will be introduced in the following sections.

1.2.1 Template Mimicking

Template-mimicking method is a simple, effective, and fast technique to construct underwater superoleophobic surface by replicating surface structures of the template. The templates can be natural surfaces,[5,20] and other artificial patterned structure.[24,25] The template-mimicking process usually contains three steps: preparing a featured template master, molding the replica, and removing the templates. Herein, we will focus on one natural surface template (lotus leaf) and the other artificial-patterned structure (micropatterned Si pillar) used as the template mimicking of underwater superoleophobic surface.

1.2.1.1 Mimicking the lotus leaf

As shown above, the lotus leaf shows feature of Janus interface, including superhydrophobicity on upper side and underwater superoleophobicity on lower side. The polydimethysiloxane (PDMS) and hydrophilic epoxy resin are chosen to replicate the upper side and lower side of lotus leaf through two-step, respectively. Generally, the epoxy resin is applied in many fields including coatings, adhesives, and composite materials due to its excellent adhesion, chemical and heat resistance, good-to-excellent mechanical properties, and very good electrical insulating properties. Selecting the epoxy resin to replicate the lower side of lotus leaf offers the artificial underwater superoleophobic surface excellent mechanical properties. Compared with the lower side of lotus leaf, this artificial superoleophobic interface materials have more excellent mechanical strength and good resistance to chemicals in water.

The whole replicating process is shown below. At temperature of 60°C, the viscosity of this epoxy resin can be kept at as low as 50 cPa·s for more than 12 h. The viscosity of the epoxy resin is low enough to completely replicate the nanoscale structure. The duplicated processing steps of the lotus leaf surface structure are shown in Fig. 1.9. A fresh natural lotus leaf floating on water surface was used as the original bio-template in the replication of step 1. The mixture of liquid PDMS and its curing agent (10:1, mass proportion) was casted onto the upper side and lower side of lotus leaf.

After solidification at room temperature for 24 h, the flexible PDMS negative replica was obtained, as shown on the left side of step 2. Then the liquid PDMS was casted on the upper side of PDMS negative template in the same manner. After PDMS solidification, the artificial upper side of lotus leaf based on PDMS was peeled off from the PDMS negative template. The liquid epoxy resin was infiltrated into the lower side of PDMS negative replica at 60°C, as shown on right side of step 2. Samples were immediately transferred to a vacuum chamber for 10 min to remove trapped air and to increase the resin infiltration through the structures. After curing, the artificial lower side of lotus leaf based on epoxy resin was released from PDMS negative replica. The artificial PDMS upper side and epoxy lower side of lotus leaf were bonded together using the oxygen plasma, as shown in step 3.

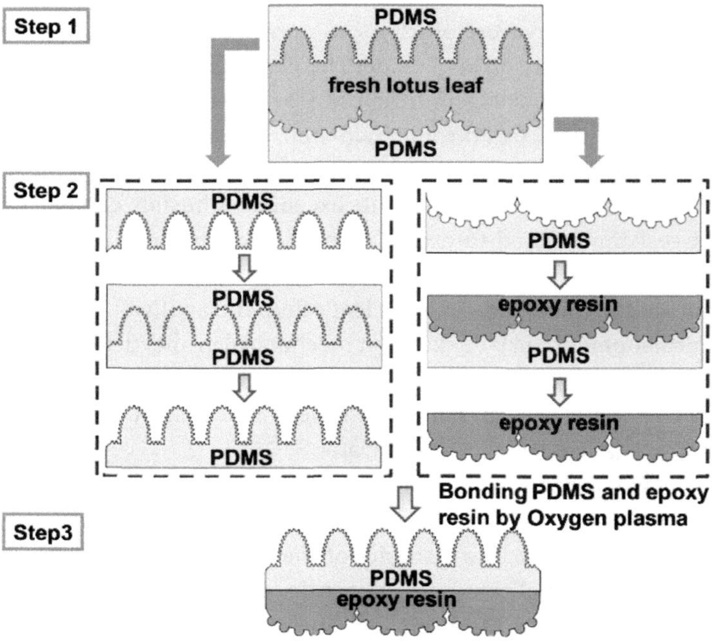

Figure 1.9 Schematic illustrating the procedure of fabricating the Janus interface materials by template mimicking.

The fabricated artificial lotus leaf floating on water surface is shown in Fig. 1.10. Its upper side shows superhydrophobicity with contact angle of 156.0 ± 1.0° in air and its lower side shows superoleophobicity with oil (*n*-hexane dyed red) of 153.5 ± 1.5° in water. The morphology of this artificial lotus leaf is shown in Fig. 1.10b and c, indicating the papillae on upper side and lower side of lotus leaf were completely replicated by PDMS and epoxy resin, respectively.

This biological surface template replicating method is more facile and fast for constructing superoleophobic surface, compared with the silicon pillar,[20] micro-patterned surface,[22] and the hybrid nanoclay hydrogels,[24] assemblied structure,[26,27] and inorganic materials.[28]

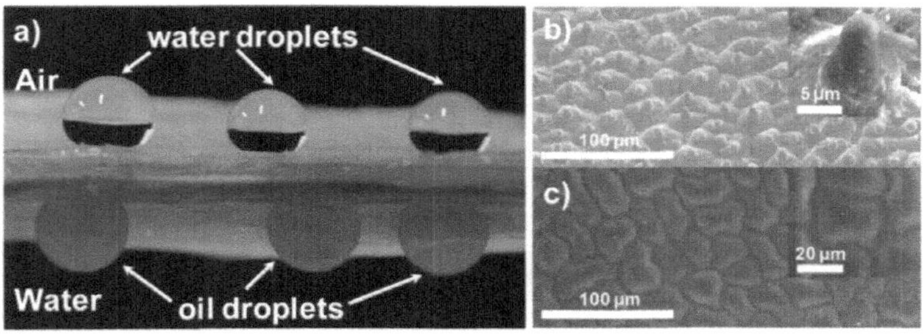

Figure 1.10 (a) The Janus interface materials is floating on the water surface, and the water droplets can stay on its upper side in the shape of sphere in air and the oil (*n*-hexane dyed red) droplets are on its lower side in the shape of sphere in water. (b) SEM images show the micro-/nanoscale hierarchical structure on the upper side of lotus leaf is replicated by PDMS. (c) The tabular and slightly convex papillae on lower side of lotus leaf are also replicated by epoxy resin.

1.2.1.2 Mimicking micropatterned surface

The other effective template method for constructing underwater superoleophobic surface is replicating the master template. The micropatterned silicon surfaces with pillars of 14 μm diameter and 30 μm with pitch value 23 μm is fabricated by photolithography.[22] Generally, to replicate this kind of micropatterned structure well, a two-step molding process was used. Firstly, a polyvinylsiloxane dental wax was applied via a dispenser on the micropatterned silicon surface and immediately pressed down by a glass plate. After complete hardening of the molding mass, the silicon micropatterned surface and the negative replica were separated. Then the negative replica was filled with a liquid epoxy resin. The specimens with negative replica were transferred to vacuum oven to remove trapped air bubbles and to assist the resin completely infiltrate the micropatterned structure. After completely curing the epoxy resin, the positive replica was separated from the negative replica. Then the underwater superoleophobic surface is obtained, and the surface morphology is shown in Fig. 1.11a. In the solid–air–water system, the CA of flat epoxy resin is only 76° and the micropatterned surface shows superhydrophobicity with CA of 155°, which is because an air pocket is formed between the pillars. However, the oil droplet is superoleophobic and OCA is about 5°. According to aforementioned mechanism of underwater superoleophic surface, when put this epoxy micropatterned structure in the water, the OCA reaches as high as 151°, which further confirmed our finding.

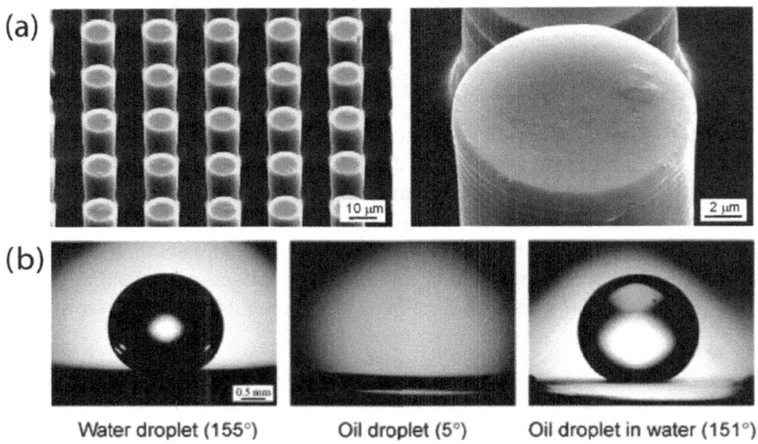

Figure 1.11 (a) SEM images show the micropatterned epoxy pillars with the diameter of 14 μm, the height of 30 μm, and the pitch of 23 μm. (b) The optical micrographs of liquid droplets on micropatterned surfaces. At a solid–air–water interface, the micropatterned surface shows superhydrophobic and superoleophilic properties, respectively. In the solid–water–oil interface system, it shows superoleophobicity with OCA of 151°.

Besides the aforementioned epoxy resin, Lin and co-workers[24] developed a bionic strategy to obtain robust underwater superoleophobic interface materials by constructing hybrid hydrogels with macromolecule and nanoclay. The robust hybrid poly(*N*-isopropylacrylamide)-nanoclay (PNIPAAm-nanoclay) hydrogels were synthesized by a photo-initiated in situ radical polymerization with clay nanoparticles as physical crosslinkers. The hierarchical structures on the hybrid hydrogel were mimicking the design-guided patterned with micro-/nanoscale topographical templates, as shown in Fig. 1.12a. The micrometer patterned protuberances are uniform with a top diameter of 71 ± 4 μm and a bottom diameter of 161 ± 2 μm. The distance between protuberances is 83 ± 3 μm. Underwater AFM image shows nanometer-scale roughness with an average root-mean-squared (RMS) roughness of 635.3 nm. The hybrid hydrogel exhibits superoleophobicity with OCA of 159.1 ± 1.6°.

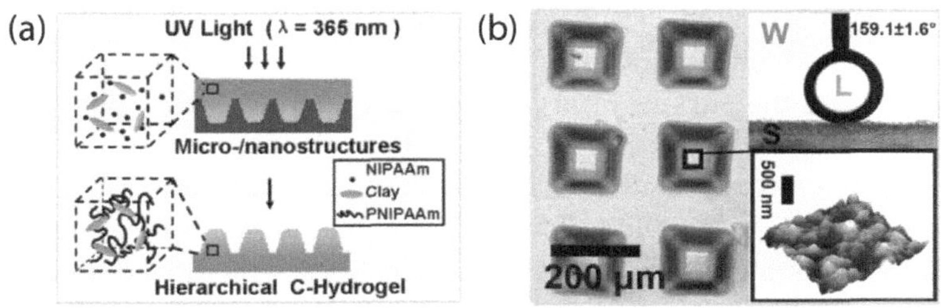

Figure 1.12 Bio-inspired strategy of constructing hierarchical PNIPAAm–nanoclay hydrogels (C-hydrogels). (a) Schematic illustration of the mold method of fabricating C-hydrogels with micro-/nanostructured surface through photo-initiated polymerization. (b) Bio-inspired hierarchical surface with microprotuberances and nanometer-scale roughness (635.3 nm RMS roughness) shows superoleophobicity.

Mechanical properties measurements show that the hybrid hydrogel can easily recover its shape when load releasing, in contrast, the as-prepared pure hydrogel breaks when press loading, as shown in Fig. 1.13. The compressive strength increases with the increase of clay content. The maximum compressive strength reaches up to 2.5 MPa at 75% strain.

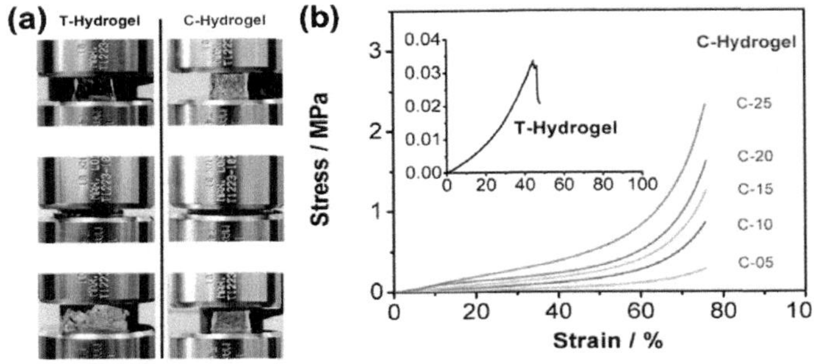

Figure 1.13 (a) Photographs of as-prepared traditional PNIPAAm hydrogel (T-hydrogel) and hybrid PNIPAAm–nanoclay hydrogel (C-hydrogel) during a compression test. The T-hydrogel easily breaks in the compress test, while C-hydrogel recovers its original shape when load releasing. (b) Stress–strain curves of C-hydrogels and T-hydrogel (inset), indicating the compressive strength increase with the increase of clay content.

PDMS is a kind of economic, biocompatible, and hydrophobic polymer, which has been widely used as interface materials by replicating the surface of organism, such as lotus leaf.[29] However, PDMS has relatively weak underwater oleophobic ability and cannot be directly used as the anti-oil application in water. Wu and co-workers[30] recently developed a facile method of fabricating the underwater superoleophobic surface of PDMS modification with oxygen plasma etching.[31] Only oxygen plasma treatment of PDMS is not enough to achieve underwater superoleophobicity because the hierarchical surface microstructures play important role. Thus, inspired by the hierarchical micro-/nanoscale of gecko foot, the gecko foot-like hierarchical structure was produced as shown in Fig. 1.14. Firstly, the hole template of silicon was created by laser interference and photolithography. Then, the PDMS was transferred to the hole template and peeled off. Finally, the gecko foot-like hierarchical structure is oxygen plasma treated. The PDMS micropillar arrays with 20 μm and 2.5 μm pitch show superhydrophilic and superoleophilic in the air environment. In water, the OCAs are 150 ± 2° and 164 ± 4°. When combining the above two types of pillar structures together, the gecko foot-like hierarchical microstructures were obtained. This surface shows underwater superoleophobcity with OCA as large as 175 ± 3° and the oil droplet easily rolls off from surface.

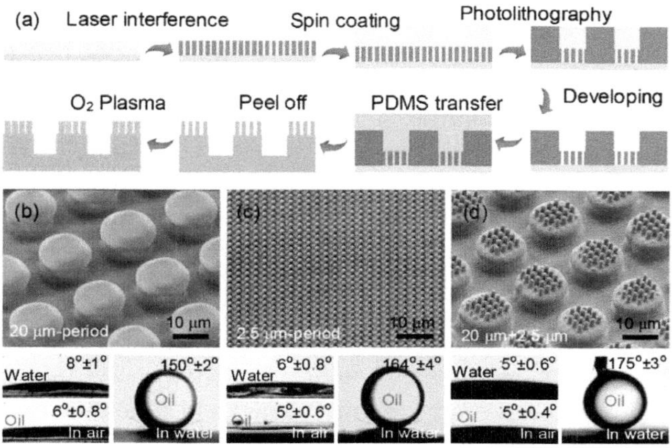

Figure 1.14 (a) The schematic procedure of fabricating underwater superoleophobic surface with PDMS through constructing gecko foot-like arrays: laser interference, photolithography, and PDMS transfer. (b–d) SEM images of 20 μm period, 2.5 μm period pillar arrays, and hierarchical gecko foot-like microstructures, and corresponding to the water and oil contact angle measurements.

This kind of underwater superoleophic PDMS materials exhibit many advantages compared to other underwater superoleophobic surface materials: more cost effective than silicon pillar,[20] better biocompatibility than epoxy resin[22] and conducting polymer,[32] and higher stability to chemical solvents than hydrogel.[24,25]

1.2.2 Surface Coating

Although replication mimicking is a kind of simple and effective approach for constructing underwater superoleophobic surface, it is very difficult to be scaled up for fabricating large-size superoleophic surface in water. Recently, Chen and co-workers have developed a facile and feasible approach—surface coating, including coating hydrogel on the hierarchical micro-/nanoscale substrate structure,[33] or stainless steel mesh,[34] and growing aligned ZnO nanorod array on the stainless steel mesh.[35]

Hydrogel is a cross-linked network with water filling the interstitial spaces, showing hydrophilic property. Herein, two typical kinds of hydrogels are used as surface coating materials to construct the underwater superoleophobic surface. The first one is poly(*N*-isopropylacrylamide) (PNIPAAm), which is a well-known thermally responsive hydrogel and shows an extended hydrophilic chain conformation below its lower critical solution temperature (LCST) in aqueous solution and undergoes a phase transition to insoluble and hydrophobic aggregates above the LCST. The typical nanoscale topography structure was successfully constructed by grafting PNIPAAm on the silicon nanowire arrays (SiNWAs) through surface-initiated atom transfer radical polymerization (SI-ATRP).[33] SEM and TEM images confirmed that the PNIPAAm was grafted on the surface of SiNWA, as shown in Fig. 1.15. And, the OCAs indicate the PNIPAAm grafted SiNWA is underwater superoleophobic below and above LCST. At the same time, the PNIPAAm-grafted SiNWA shows a low adhesive force (about 4.6 μN) and almost no residuum to oil droplets. This underwater superoleophobic PNIPAAm-grafted SiNWA showed largely reduced platelet adhesion both below and above the LCST of PNIPAAm.

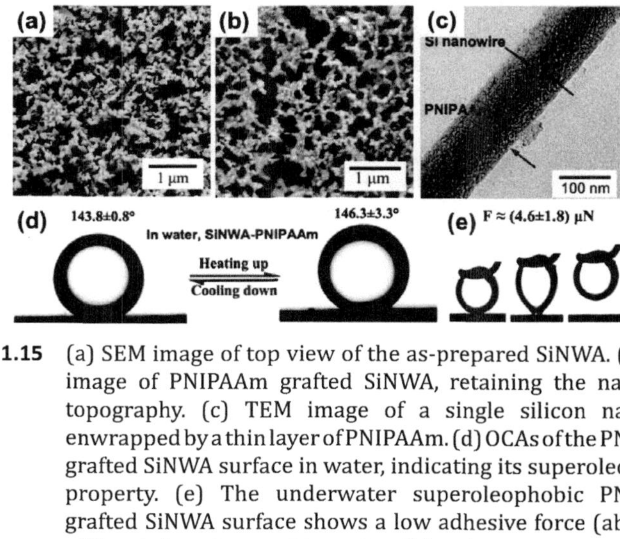

Figure 1.15 (a) SEM image of top view of the as-prepared SiNWA. (b) SEM image of PNIPAAm grafted SiNWA, retaining the nanoscale topography. (c) TEM image of a single silicon nanowire enwrapped by a thin layer of PNIPAAm. (d) OCAs of the PNIPAAm grafted SiNWA surface in water, indicating its superoleophobic property. (e) The underwater superoleophobic PNIPAAm grafted SiNWA surface shows a low adhesive force (about 4.6 μN) and almost no residuum to oil droplets.

The other typical and environmentally friendly hydrogel material is polyacrylamide (PAM), which was chosen as the coating material to fabricate underwater superoleophic surface.[34] Stainless steel meshes were used as the substrates. After cleaning, the stainless steel mesh was immersed in the mixed pre-gel solution, and then the in situ redical polymerization was initiated by UV. The obtained PAM hydrogel-coated mesh is shown in Fig. 16. SEM images show that the PAM hydrogel has been successfully coated on the stainless steel mesh. The higher magnification SEM image indicates the surface of PAM hydrogel consists of numerous nanostrucated papillae. The OCA of PAM hydrogel coated mesh reaches as high as 155.3 ± 1.8°. And, the dynamic underwater oil-adhesion measurements revealed that the PAM hydrogel-coated stainless steel mesh exhibits excellent ultralow adhesion to the oil droplet and no oil residuum on its surface. This novel superhydrophilic in the air and underwater superoleophobic hydrogel-coated microscale porous metal mesh in an oil/water/solid three-phase system, shows promising application in oil/water separation, which will be introduced in details in section 1.3.2.

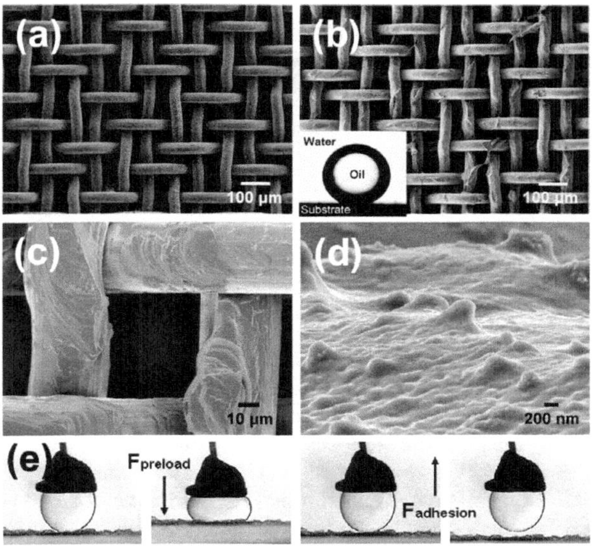

Figure 1.16 (a) SEM image of uncoated stainless steel mesh. (b) The PAM hydrogel-coated stainless steel mesh shows underwater superoleophobic property with OCA of 155.3° ± 1.8°. (c) Enlarged SEM view of a single pore of the PAM hydrogel-coated stainless steel mesh. (d) The higher magnification SEM image of one single wire on hydrogel-coated stainless steel mesh, on which the nanoscale papillae can be clearly observed. (e) Photographs of the dynamic underwater oil-adhesion measurements on the PAM hydrogel-coated stainless steel mesh. The coated meshes exhibit excellent ultralow adhesion to the oil droplet and no oil residuum on the coated mesh.

Recently, Tian and co-workers[35] have developed another approach of surface coating to construct the underwater superoleophobic surface through growing an aligned ZnO nanorod array on the stainless steel mesh through two steps, as shown in Fig. 1.17. SEM images show that the nanoscale morphology of the coated mesh and results show underwater superoleophobic property with OCA of 156°. Due to the photo-induced property of ZnO array, this kind of ZnO-coated steel mesh can be applied as photo-induced water–oil separation film, which will be introduced in details in section 1.3.2.

Figure 1.17 (a) Schematic illustration of the preparation process: firstly, ZnO crystal seeds were prepared by ZnO sol dip-coating onto the stainless steel mesh substrates and annealed. Then the as-prepared substrates were grown in an aqueous solution of zinc nitrate hydrate and methenamine. Finally, they were rinsed with deionized water and dried. The inner circularity represents the stainless steel wire and the outer radial pattern is the corresponding aligned ZnO nanorod array layer. (b) SEM image of coated mesh film. (c) SEM image of side view of the aligned ZnO nanorod array-coated stainless steel mesh film. (d) OCA of the coated mesh is about 156° in water, indicating its superoleophobic property.

1.2.3 Assembly

The aforementioned surface coating for constructing the underwater superoleophoic interface materials is very simple. However, these underwater superoleophobic coatings may be not stable in complex marine environment, especially in high ionic strength seawater.

Recently, Xu and co-workers[26] have successfully fabricated a stable underwater superoleophobic organic/inorganic hybrid interface materials through layer-by-layer (LBL) assembling polyelectrolytes and citrate-capped Au nanoparticles (cit-AuNPs). LBL assembly is a versatile and cost-effective approach, which can combine different functions into a single surface. The whole assembly process is shown in Fig. 1.18. To completely simulate the salinity of seawater, the high-ion-strength condition of 0.5 M NaCl was chosen to perform the LBL process. The glass substrates were firstly coated with four double layers of poly(diallyldimethlammonium chloride) (PDDA) and poly(4-styrenesulfonic acid) (PSS). Then seven bilayers of cit-AuNPs and PDDA were deposited on PDDA/PSS LBL films.

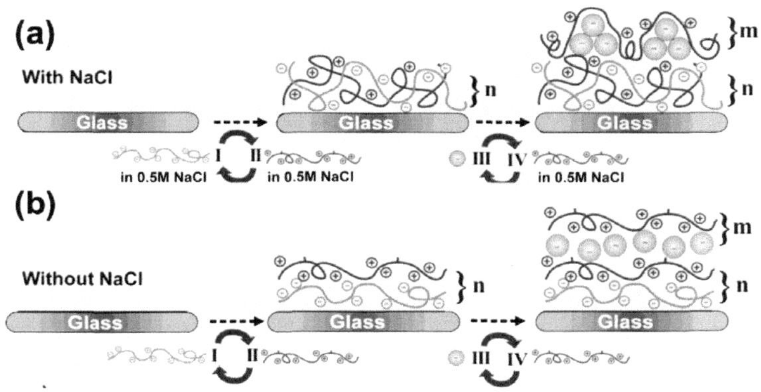

Figure 1.18 Layer-by-layer schematic process to fabricate organic/inorganic hybrid underwater superoleophic and low-adhesive interface materials. (a) Fabrication of $(AuNPs/PDDA)_{7\text{-salt}}$ surface in 0.5 M NaCl; (b) fabrication of $(AuNPs/PDDA)_7$ without adding NaCl.

The hybrid LBL film $(AuNPs/PDDA)_7$ showed oleophobicity with an OCA of 128.0 ± 3.2° underwater, and the LBL film $(AuNPs/PDDA)_{7\text{-salt}}$ exhibits superoleophobicity with OCA of 168.3 ± 0.6° and low oil-adhesion force of only 2.0 ± 4.4 μN. Obviously, these underwater oil-repellent properties of LBL hybrid interface materials are not only contributed to the used chemical composition, but also to the surface roughness. The atomic force microscope (AFM) images of $(AuNPs/PDDA)_7$ and $(AuNPs/PDDA)_{7\text{-salt}}$ were shown in Fig. 1.19. On $(AuNPs/PDDA)_7$ film, the individual AuNPs were dispersed randomly and its RMS roughness is 14.4 nm and the maximum height fluctuation is 57 nm. After being immersed in 0.5 NaCl solution for one day, the RMS roughness increases to 15.1 nm due to some cit-AuNPs aggregation. While for $(AuNPs/PDDA)_{7\text{-salt}}$, large aggregations of cit-AuNPs were embedded in the film, resulting in the increase of RMS roughness of 17.1 nm. And, the maximum height fluctuation increases up to 95 nm. These results indicate that NaCl can lead to high surface roughness.

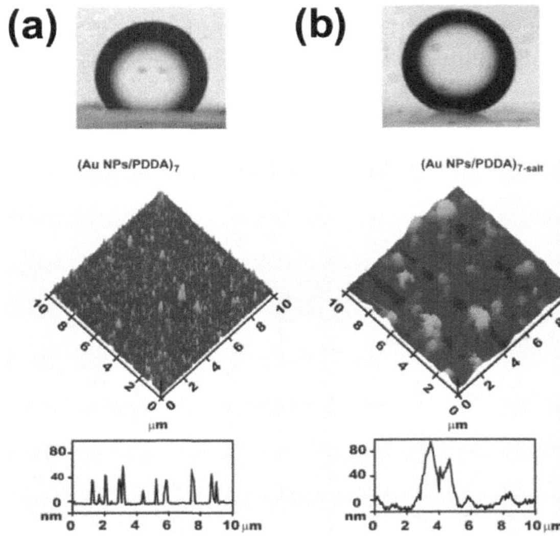

Figure 1.19 AFM images of underwater superoleophoic interface (a) AFM image shows the surface of $(AuNPs/PDDA)_7$ is relatively smooth and underwater oleophicity with OCA of 128.0 ± 3.2°. (b) After introducing ions into polyelectrolyte solution, the as-prepared $(AuNPs/PDDA)_{7\text{-salt}}$ turns much rougher and fluctuation range is larger and shows underwater superoloephicity with OCA of 168.3 ± 0.6°.

On the other hand, we also improved the other easy-operation and cost-effective assembly way to construct the underwater superoleophobic organic/inorganic nacre-like composite film.[27] The layered poly(*N*-isopropylacrylamide)-nanoclay (PNIPAM-nanoclay) hydrogel composite film was fabricated, as shown in Fig. 1.20a. The nanoclay platelets, the PNIPAM monomer, and the initiator were assembled through vacuum filtration. The UV light was used to initiate the in situ radical polymerization, and then the nacre-like organic/inorganic hybrid composite film was obtained, which shows the underwater superoleophocity with OCA of 160.5 ± 2.3°. The AFM image shows the typical hierarchical micro-/nanoscale structure on the surface under water, which plays a great role in its underwater superoleophobicity. This kind of nacre-like underwater superoleophic hybrid composite film shows the higher mechanical properties comparied with the mimicking the micropatterned surface using hydrogel[24] and PDMS.[30]

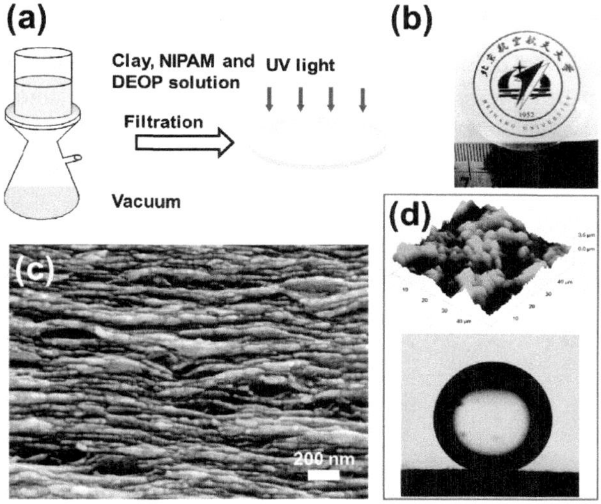

Figure 1.20 (a) Vacuum-assisted infiltration process for fabricating nacre-like organic/inorganic composite film. (b) Digital photograph of a free-standing transparent underwater superoleophobic hybrid film. (c) Cross-section morphology of a dried nacre-like hydrogel film with typical layered structure. (d) AFM image shows the hierarchical micro-/nanoscale structure of hybrid surface and underwater superoleophicity with OCA of 160.5 ± 2.3°.

1.2.4 Inorganic Materials

Inspired by the underwater low adhesive superoleophobicity of the pallium-covered region of a short clam's shell, Liu and co-workers successfully fabricated a novel artificial underwater superoleophobic high-energy coating based on copper oxide (CuO).[28] The copper sheets were firstly immersed in aquous ammonia (pH of 10.5) for

24 h at room temperature. The color of these sheets changed from bright yellow to dark black, indicating the formation of copper oxide films. SEM images reveal that the surface morphology of black film is micrometer-scaled cauliflower-like clusters with nanometer-scaled branches, as shown in Fig. 1.21b. A smooth CuO coating fabricated as the control for comparison shows underwater oleophobic property with OCA of 135.5 ± 3.8°. The hierarchical CuO coating shows underwater superoleophobicity with OCA of 163.0 ± 4.5°. The oil droplet adhered on the smooth CuO with highly adhesive force of 72.2 ± 6.8 μN, but easily rolled off the hierarchical CuO when tilted. The adhesion of oil on this kind of coating was too low to be detected (less than 1 μN), indicating the hierarchical cauliflower-like micro-/nanoscale structure has underwater superoleophobicity with low oil adhesion.

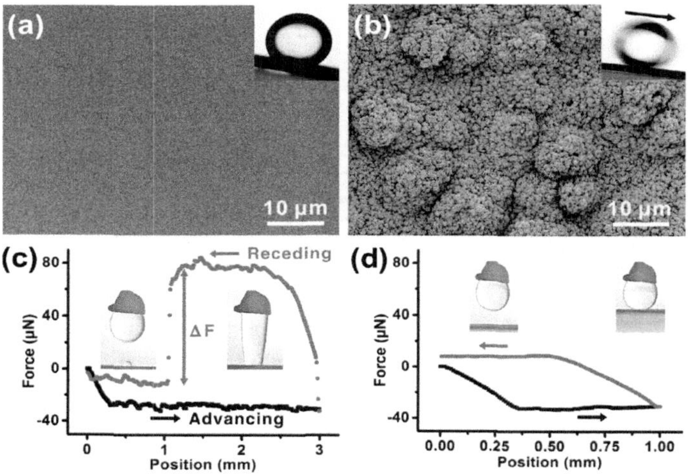

Figure 1.21 (a & b) SEM images of smooth CuO and cauliflower-like CuO surface. (c & d) The measurement of underwater oil adhesion on these two coatings.

This kind of bio-inspired functional coating can be widely extended to other inorganic material systems, providing a clue to the fabrication of underwater superoleophobic engineering metal surfaces for protecting aquatic equipment from oil contamination. Recently, Nosonovsky and co-workers developed a much simpler method for constructing underwater superoleophobic surface based on strong Al-based alloys[36] and corrosion-resistant brass.[37] For example, the brass was mechanically abraded with 50 grit silicon carbide papers.

1.3 Promising Applications

Underwater superoleophobicity has broad promising applications, such as external stimuli-responsive surface, oil/water separation, anti-fouling microfulidic channel, and oil capture under water. Herein, these applications will be discussed in details in the following sections.

1.3.1 Underwater Smart Surface

As discussed before, the underwater functional surfaces have been investigated and bio-inspired synthetic methods have been developed. Recently, numerous attentions have been focused on the construction of the stimuli-responsive surface, or smart surface, which shows switchable wettability and controllable adhesion for oils under water.[38] This kind of underwater smart surface has a great application in many fields, such as sensors, microfluidic devices, lab-on-chip fields, etc. Many underwater smart surfaces driven by external stimuli, including temperature,[25,33] pH,[39,40] and electrical potential,[32,41,42] have been successfully fabricated. In this section, we will mainly discuss underwater smart surfaces with switchable wettability and controllable adhesion for oils.

1.3.1.1 Thermal responsive surface

Recently, Chen and co-workers[25] investigated a thermal-responsive poly(*N*-isopropylacrylamide) (PNIPAM) hydrogel at the water/solid interface. The PNIPAM hydrogel surface shows reversible switching between a superoleophobic and low oil adhesion below PNIPAM lower critical solution temperature (LCST, ~32°C) and an oleophobic and high oil adhesion above its LCST. As shown in Fig. 22, the oil droplet (1,2-dichloroethane) could easily roll off the PNIPAM hydrogel surface when temperature is below its LCST, showing a low oil adhesion. When the temperature increases above its LCST, the oil droplet firmly adhered on its surface, even if the surface was turned upside down. If using another kind of oil droplet (n-hexane) on its surface at 40°C, and then cooled down the temperature to 23°C, the n-hexane droplet can fly away from its surface. At 23°C, the recorded maximum oil adhesion force is only 5.8 ± 1.8 μN. In contrast, when temperature increases over its LCST, for example at 40°C, the maximum oil adhesion force reaches as high as 23.1 ± 1.9 μN. In fact, according to measurement process of the oil adhesion force, the oil droplet always has been greatly distorted and finally broken into two parts, indicating the actual adhesion force should be even higher than the recorded maximum oil adhesion force.

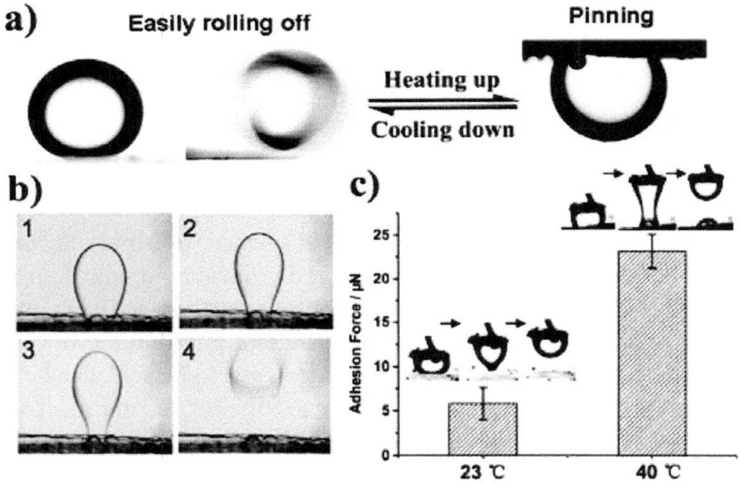

Figure 1.22 Demonstration of the thermal-responsive oil adhesion on the PNIPAM hydrogel surface underwater. (a) The oil droplet can easily roll off at 23°C, while firmly adheres on the surface even if the PNIPAM hydrogel surface is placed upside down at 40°C. (b) An n-hexane oil droplet flew away when temperature decreasing under water. (c) The adhesive force is 5.8 ± 1.8 μN at 23°C and 23.1 ± 1.9 μN at 40°C. The inset images are measurements process (from left to right): the oil droplet is contacting the surface, the oil droplet is leaving the surface, and the oil droplet has left the surface.

The mechanism of thermal-responsive PNIPAM hydrogel surface was investigated with in situ AFM, as show in Fig. 1.23. At 23°C, the intermolecular hydrogen bonds between hydrophilic groups and water molecules were formed, leading to a hydrated, swollen state of the PNIPAM hydrogel. Thus, the surface exhibits low roughness and high ratio of water content. Therefore, the superoleophobic state and low oil adhesion between water and oil was achieved on the PNIPAM hydrogel surface. In contrast, when temperature rises to 40°C over its LCST, the intramolecular hydrogen bonds between neighboring PNIPAM branched chains were formed, which leads to a dehydrated collapsed state and thus offers a surface with enhanced roughness and a lower ratio of water content. Thus, the thermal-responsive surface with oleophobic state and higher oil adhesion was obtained.

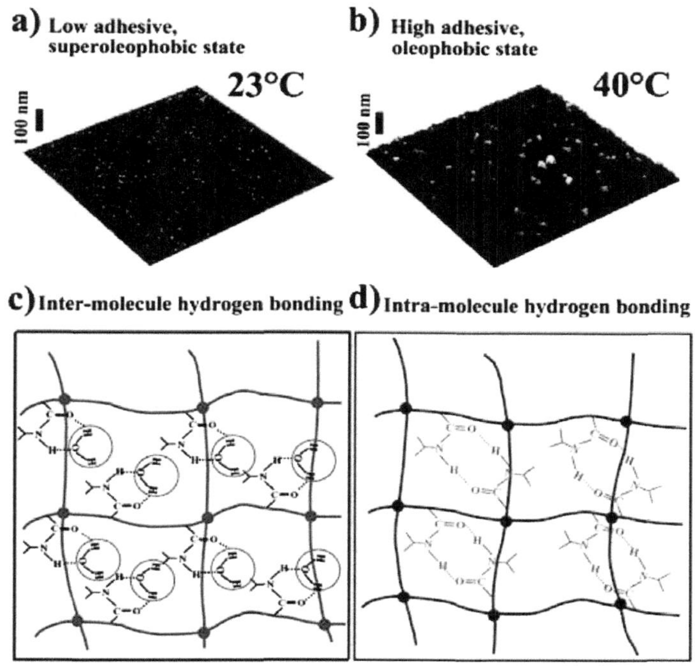

Figure 1.23 The atomic force microscopy (AFM) images of PNIPAM hydrogel surface under water and proposed thermal responsive mechanisms. (a) Quite smooth topography at 23°C. (b) Greatly enhanced roughness at 40°C under water. (c) At low temperature, the intermolecular hydrogen bonds were formed between PNIPAM and water molecules, keeping the PNIPAM hydrogel surface with a high ratio of water content. (d) At high temperature, the intramolecular hydrogen bonds were formed in the neighboring PNIPAM chains, resulting in lower ratio of water content.

1.3.1.2 pH responsive surface

Cheng and co-workers[39] reported a pH-responsive underwater superoleophobic surface, which is fabricated by plasma-induced grafting the poly(acrylic acid) (PAA)[43] on glass substrate. PAA is a typical pH-responsive polymer, which has been wildly used to construct the pH-responsive wetting surface[44] and pH-gating ion nanochannel.[45] The pH-responsive underwater superoleophobic surface is fabricated by grafting PAA nanostructured film onto a glass substrate (PAA-G) through a plasma polymerization[46] process, as shown in Fig. 1.24. The linear PAA chains were grafted on the glass surface and the thickness of PAA, calculated by AFM, was about 64.5 ± 9.2 nm. The surface of the PAA-G in solution with a pH value of 2.4 exhibits rough morphology as shown by the underwater AFM image (Fig. 1.23b). The surface morphology becomes smooth when immersing PAA-G into a solution with a pH value of 8.0 (Fig. 1.23c). The blue dash rectangle in Figs. 1.24b and c is used to mark the same location in the in situ processing of underwater AFM testing. The blue dash rectangle mark proves the in situ transition of surface morphology on PAA-G when the pH value of solution is greater than the pKa of PAA. The surface roughness decreases from about 5 nm to 1 nm, when the pH value changes from 2.4 to 8.0. The carboxylic pendant groups of PAA accept protons at a pH value lower its pKa of 4.7 (insert in Fig. 1.24a),[47] release them with an increase of pH value and transform into polyelectrolytes due to electrostatic repulsion forces of the molecular chains.[48] The conformation of PAA changes from coiled at low pH to stretched at high pH,[49] schematically outlined in Fig. 23a (lower insert). At low pH, the formation of intramolecular hydrogen bonds prevail in PAA films, while at high pH intermolecular hydrogen bonds were formed between surrounding water molecules and PAA instead of intramolecular hydrogen bonds.

The underwater oil-adhesion shows responsive to pH, due to the morphology transition of PAA at the pKa of 4.7. The oil-adhesion dramatically decreases with pH values increasing from 4.6 to 5.0. When the pH value is below 4.7, the PAA-G exhibits high adhesion to the oil droplet, and always draws oil droplets with a certain distortion. However, there is no oil residuum on the PAA-G when the oil droplet leaves. At a pH value of 1.0, the recorded adhesive force reaches up to 21.6 ± 5.0 μN. With the pH values increasing from 1.0 to 4.6, the adhesive force decreases to 15.0 μN from 2.9 μN. Further, increasing pH values above the pKa of PAA, the oil-adhesion dramatically decrease to 2.0 ± 0.6 μN with a pH value of 5.0, 1.5 ±

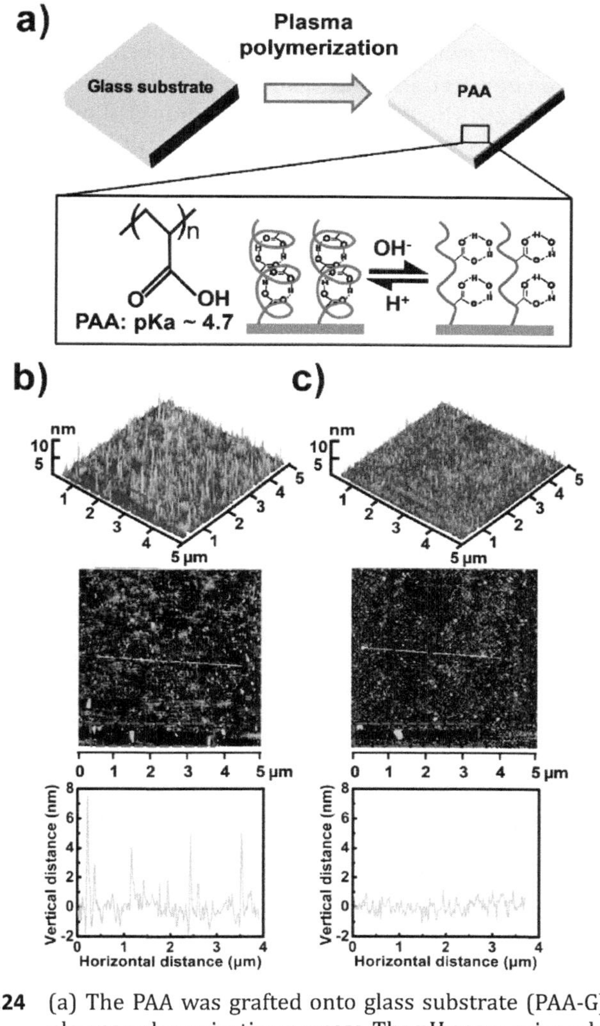

Figure 1.24 (a) The PAA was grafted onto glass substrate (PAA-G) by the plasma polymerization process. The pH-responsive schematics of the conformation switch of PAA: the intramolecular hydrogen bonds form among the carboxylic acid groups on the PAA chains when the pH value is below pKa, and the intermolecular hydrogen bonds form between PAA chains and water molecules when the pH is above the pKa. (b) Underwater AFM image of PAA-G in the solution with a pH value of 2.4 shows rough morphology with a roughness of about 5 nm. (c) When PAA-G is in solution with pH value of 8.0, the in situ AFM image becomes smooth with a roughness of about 1 nm.

0.4 μN with a pH value of 8.0 and zero with a pH value of 12.0. In addition, the underwater in situ transition experiment from low to high adhesion between the PAA-G and oil droplet was carried out, as shown in Fig. 1.25. The PAA-G was firstly immersed into 0.01 mol/L sodium hydroxide (NaOH, pH = 12.0) solution and an oil droplet was dropped onto its surface keeping a spherical shape. Then, 0.1 mol/L hydrochloric acid (HCl, pH = 1.0) was quickly added, resulting in the pH value quickly switching from 12.0 to 1.0. The adhesive force switches from low to high in around 1 s. Therefore, the adhesive force can be easily tuned between high and low by adjusting the environmental pH values. To investigate the reversibility of the pH-responsive oil-adhesion switch, the adhesive force of PAA-G in solutions with pH values of 2.4 and 8.0 was alternately tested. The measurements of adhesive force from 10 consecutive cycles (Fig. 1.25b) show a good reversibility of this switchable PAA-G surface between high and low oil-adhesion responsive to pH alternation.

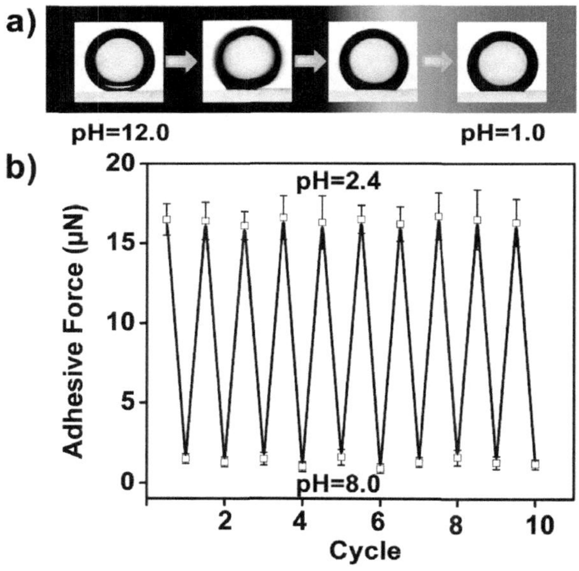

Figure 1.25 (a) The in situ process of an oil droplet on PAA-G changes from low to high adhesive force by quickly changing the environmental pH value from 12.0 to 1.0. (b) The adhesive force decreases from 16.7 ± 2.1 μN to 1.5 ± 0.4 μN when pH increases from 2.4 to 8.0. The adhesive force can easily switch for 10 cycle times between a pH value of 2.4 and pH value of 8.0, indicating a good reversibility.

On the other hand, to further verify the pH-responsive switchable oil-adhesion on the PAA-G, other oils with different surface tensions were also utilized. The adhesive forces between PAA-G and other oils such as hexane (18.0 mN/m), chloroform (28.9 mN/m), tetrachloroethane (33.9 mN/m), and dichlorobenzene (36.7 mN/m) were also tested. The comparison results show that all adhesive forces dramatically decrease when pH value is greater than the PAA pKa of 4.7. This smart surface can reversibly switch between high oil-adhesion at low pH and low oil-adhesion at high pH underwater along with the pH-responsive transition of molecular conformation of PAA. It supplies a promising platform for tuning surface properties and has potential applications in smart microfluidics, controllable bio-adhesion, and intelligent materials for oil-removal treatment and marine anti-fouling.

Recently, Liu and co-workers[40] reported a novel approach to tune the underwater surface wettability based on the Lewis acid–base (AB) interactions, which is electron-acceptor and electron-donor interactions. The electron donor–accepter interactions result in attractive force, which can greatly decrease the interfacial tension.[50]

The 1,2-dichloroethane and hexane that were used as polar and apolar oils, respectively, and the aqueous solutions with pH values from 2 to 13 were used as tunable aqueous phases from acidic to basic. As confirmed before, the microstructures can greatly enhance the surface wettability. The substrate is oleophilic for 1,2-dichloroethane in acidic aqueous phase (pH 3). When the aqueous phase changes to basic (pH 12), the OCA reaches as high as 150°, indicating surface becomes superoleophobic (Fig. 1.26a). This is because the trapped basic aqueous solution in the microstructures contacts with a Lewis acidic oil, and Lewis AB interactions at the oil–water interface happen. In contrast, no matter whether the aqueous phase is acidic or basic, the apolar hexane drop always spreads on the microstructured surface due to the absence of Lewis AB interactions at the oil–water interface (Fig. 1.26b).

Different from the above pH-responsive underwater wettability controllable approach, this strategy for tunable surface wettability is independent of the solid phase. The underwater wettability can be tuned through control of the Lewis AB interactions at the liquid–liquid interface.

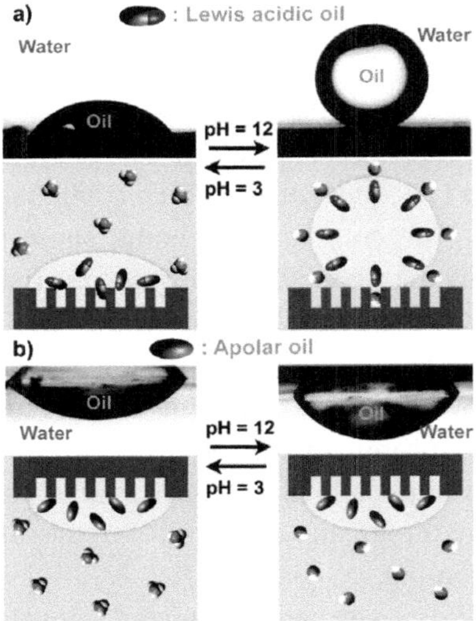

Figure 1.26 (a) For the Lewis acidic oil (1,2-dichloroethane), an oleophilic-to-superoleophobic transition is observed. (b) For the apolar oil (hexane), the surface remains oleophilic because of the absence of Lewis AB interactions.

1.3.1.3 Electrochemical responsive surface

The aforementioned stimuli-responsive surfaces are required to ex situ change underwater wettability after the surface were given the external stimuli; in other words, the in situ underwater wettability switch is not achieved. Liu and co-workers[32] developed a simple and efficient strategy to achieve an in situ underwater wettability switch through electrochemical potential. The polypyrrole (PPy) film was chosen as the substrate, and its surface composition and micro-/nanostructure can be electrochemically tuned, resulting in that the wetting state of oil droplet was reversibly switched between partly wetting and super–anti-wetting state. This reversibly underwater wettability switched PPy film can be applied for smart control of oil droplet's motion, by switching the electrochemical potential of the film, as shown in Fig. 1.27. The PPy film was tilted around 4°. An oil droplet (1,2-dichloroethane) parked and adhered on PPy film at a positive voltage (0.55 V vs. Ag/AgCl). When negative voltage (–0.9 V vs. Ag/AgCl), the PPy film became superoleopobic, and the oil droplet started to roll down. And, the cycle can be easily repeated through switching the positive voltage (0.55 V vs. Ag/AgCl) and the negative voltage (–0.9 V vs. Ag/AgCl). Therefore, the motion of the oil droplet can be controlled by in situ tuning the wettability of the PPy film.

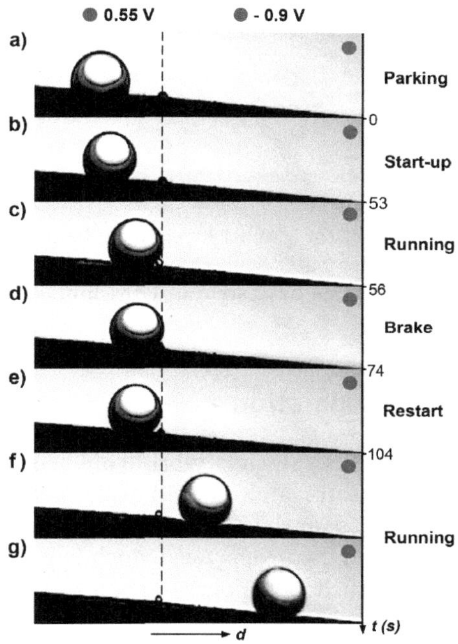

Figure 1.27 The smart control of an oil droplet motion on the PPy film surface can be repeated through switching the electrochemical potentials. (a) An oil droplet (1,2-dichloroethane) parked on the PPy film surface. (b) The OCA increased when the PPy film was reduced. (c) The oil droplet began to run on the surface. (d) The oil droplet braked and adhered on the PPy film again. (e–g) The oil droplet came back to the ball state and rolled down.

The electrochemical switching can also be applied to design mechanical hand for oil droplet transportation without loss under water. Ding and co-workers[42] fabricated aligned polyaniline (PANI) nanowire film with reversible superoleophobic property with electrochemical switching. The transportation process of an oil droptlet from one PANI film to another is controlled by electrical potential, as demonstrated in Fig. 1.28. Firstly, an oil droplet (1,2-dichloroethane) was put on the aligned PANI nanowire film (F1) without applied electrical potential. Secondly, another aligned PANI nanowire film (F2) with applied a voltage of –0.2 V, moved slowly towards F1, and then the oil droplet completely adhered on the F2 surface. The first transportation of oil droplet is finished. Following the same procedure, the oil droplet can be transferred back to F1 by applying a voltage of −0.2 V.

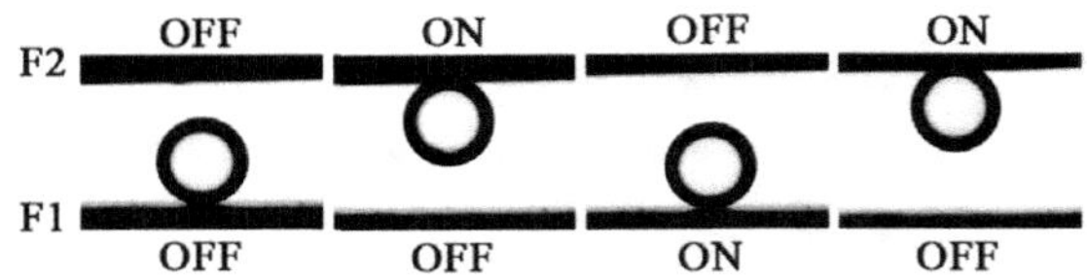

Figure 1.28 The snapshots of a smart mechanical hand transferring an oil droplet under water from emeraldine salt state of PANI film to leucoemeraldine base state of PANI film applied potential of 0.2 V.

1.3.2 Oil/Water Separation

Oil/water separation is always worldwide challenge due to industrial oily waste water and oil spill accidents. The traditional oil/water separation materials are developed based on their hydrophobic and oleophilic properties, which is usually called "oil-removing" materials, such as kapok,[51] carbon-based materials,[52–56] hydrophobic aerogels,[57,58] polytetrafl uoroethylene (PTFE) coating mesh,[59] polydimethysiloxane (PDMS) coated nanowire membrane,[60] nanoporous polydivinylbenzene materials,[61] crosslinked oil-absorbing polymer gels,[62–64] and the fluorinated-modified polyurethane (PU)-foam.[65] However, these oil/water separation materials have some limitations such as easily fouled and difficult recyclable,[23,66,67] especially for the high-viscosity oils.

Recently, the novel type of oil/water separation materials based on the superhydrophilic and underwater superoleophibic hydrogel was successfully developed by Jiang's group.[34,68] Compared to the traditional "oil-removing" oil/water separation materials, this kind of oil/water separation can be called "water-removing" materials, which overcomes the easy-fouling and hard-recycling limitations in essence. It is a new attempt to design next-generation oil/water separation based on the underwater superoleophobicity, suggesting attractive applications in oily waste water treatments and oil spill cleanup. The fabrication process of "water-removing" oil/water separation materials has been described in section 1.2.2. Figure 1.29 illustrates the oil/water separation experiment procedure, separation efficiency, and the intrusion pressure this "water-removing" oil/water separation material supported. The mixture of crude oil and water with ratio of 30:70 in volume was poured onto the separation coated-mesh. Crude oil was repelled and kept in the upper glass tube, and the water quickly permeated and dropped into the glass beaker below. There is no external force applied in the whole separation process. This kind of novel "water-removing" oil/water separation materials can be used to separate many organic solvent/water mixtures, including gasoline, diesel, vegetable oil, hexane, and petroleum ether, as shown in Fig. 1.29c. And, the separation efficiency is above 99%. On the other hand, the intrusion pressure of oils can be calculated with the density of used oils. The average intrusion pressure for the chosen oils was above 1.0 kPa, as shown in Fig. 1.29d.

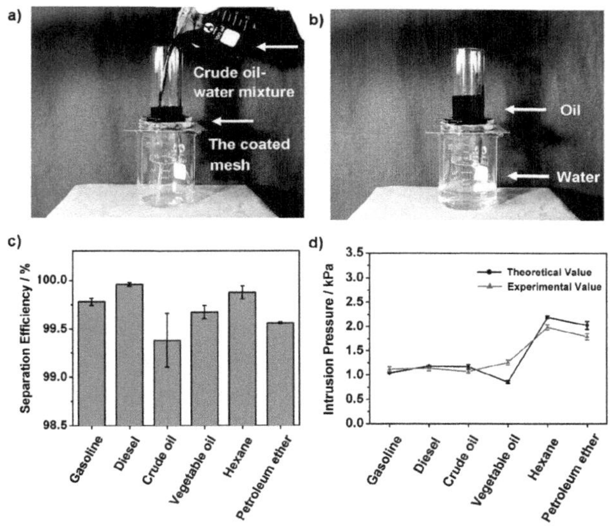

Figure 1.29 Demonstration of novel "water-removing" oil/water separation materials based on PAM hydrogel-coated mesh with pore size of about 50 μm. (a) The oil/water separation materials were fixed between two glass tubes, the mixture of crude oil and water was poured into the upper glass tube. (b) Water in the mixture permeated through oil/water separation materials, while the crude oil was repelled and kept in the upper glass tube. (c) The separation efficiency of the novel "water-removing" oil/water separation materials for a selection of oils in terms of their oil rejection coefficient. (d) The theoretical and experimental values of intrusion pressures for a series of oils.

Although the oil/water separation materials have been widely investigated, the research on controllable oil/water separation in response to external stimuli has started. To develop the controllable oil/water separation system, a smart surface that can switch its oil wettability in response to external stimuli in aqueous media, should be firstly prepared. Although numerous efforts have focused on controlling oil wettability as well as oil-adhesion properties on material surfaces in aqueous media,[25,32,33,39–42,69] the smart surfaces with switchable superoleophobicity and superoleophilicity in aqueous media have been rarely reported.[35,70] Generally, the aforementioned methods of constructing the underwater superoleophobic surfaces combine the hydrophilic composition and hierarchical macro-/nanoscale structures. Although the underwater superoleophobicity can be easily obtained, it is not possible to switch to superoleophilicity, and this is because the hydrophilic surface traps water and repels oils. To realize underwater superoleophilicity, a surface chemistry should be both oleophilic and hydrophobic.[22,60,61,71] Therefore, to construct a smart surface with switchable oil wettability between underwater superoleophobicity and superoleophilicity, the surface chemistry should comprise both hydrophilic and oleophilic/hydrophobic characteristic in response to external stimuli. Herein, we introduce the recently developed controllable oil/water separation in responsive to external stimuli, including pH-responsive[70] and UV photo-induced.[35]

Zhang and co-workers[70] grafted a block copolymer comprising pH-responsive poly(2-vinylpyridine)(P2VP) and oleophilic/hydrophobic polydimethylsiloxane blocks (i.e., P2VP-b-PDMS), to non-woven textiles and polyurethane sponges, to prepare surfaces with switchable underwater superoleophilicity and superoleophobicity. Figure 1.30 shows the different oil-wetting behaviors of the functionalized textile immersed in water with different pH values of 6.5 and 2.0. An oil (1,2-dichloroethane) droplet sucked into the textile within 0.12 s, demonstrating its underwater superoleophibicity. And, this property can be used in underwater oil capture, as shown in Fig. 1.28a. In contrast, when immersed in water with a pH of 2.0, the oil droplet formed a sphere with a OCA of 165.3° on the textile surface, indicating its underwater superoleophobicity. The mechanism of this smart transition can be explained by the Fig. 1.30c. In water with a pH of 6.5, the P2VP is not protonated and maintains its collapsed conformation, resulting in the dominant exposure of the PDMS, which retains its high affinity to oil. With the assistance of the hierarchical surface structures, the functionalized textile shows underwater superoleophilicity. The PDMS chains stretch out into the oil due to their oleophilic nature. Moreover, the air could be trapped inside the porous structures of the textile in the process of moving superhydrophobic textile into water from air, which may also contribute to the underwater superoleophilicity. In contrast, for water with a pH of 2.0, the functionalized textile can be completely wetted with water, and the protonated P2VP chains exhibit an extended conformation and are dominantly exposed. Thus, the water-trapped surface formed on top of the oleophilic PDMS, blocks the access of the oil by the PDMS, resulting in underwater superoleophobicity. On the hand, the experiments indicate that the reversible cycle can be repeated many times without any change in the pH-responsive property of the surface.

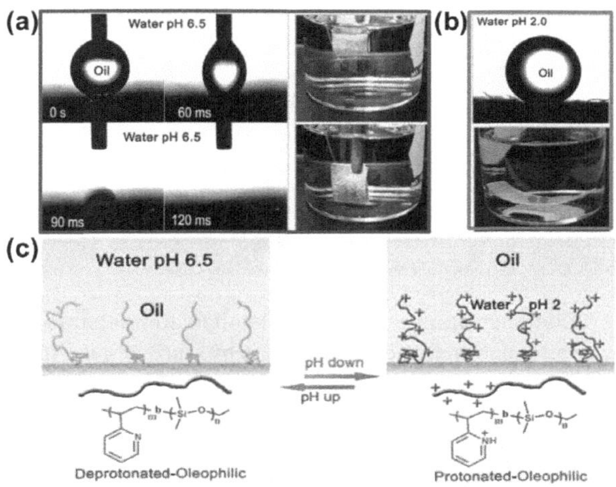

Figure 1.30 (a) OCA measurements of a 1,2-dichloroethane droplet on the functionalized textile surface in water with a pH of 6.5, and corresponding to snapshots the capture of an oil droplet in water with a pH of 6.5. (b) OCA measurements of a 1,2-dichloroethane droplet on the surface of the textile in water with a pH of 2.0, and corresponding photograph of OCA measurements of a 1,2-dichloroethane droplet droplets sitting on the functionalized textile in acidic water with a pH of 2.0. (c) Schematic diagrams for the switchable oil wettability of the P2VP-b-PDMS-grafted textile in water with a pH of 6.5 (left) and 2.0 (right).

The demonstration of controllable oil/water separation is shown in Fig. 1.31. The mixture of commercial gasoline and water with pH of 6.5 was poured into the upper glass tube. The gasoline passed through the textile, but the water stayed in the upper glass tube. When the textile was simply wetted by water with pH of 2.0, the water quickly passed through but the gasoline stayed in the upper glass tube. This kind of versatile controllable oil/water separation materials will bring a new concept to design and fabricate intelligent interfacial materials for underwater applications.

Figure 1.31 Demonstration of controllable oil/water separation based on the functionalized textile. (a) A mixture of gasoline and water (pH of 6.5) was poured into the upper glass tube. The gasoline passed through the textile, but the water stayed in the upper glass tube. (b) The functionalized textile was simply wetted with acidic water (pH of 2.0) before oil/water separation. The water can easily pass through the textile, whereas the gasoline remained in the upper glass tube.

Recently, Xue and co-workers[72] developed a simple method to realize the in situ wettability switch at different temperatures, which is much more simpler than the previous report.[70] The block copolymer (BCP) of PMMA-b–PNIPAAm was chosen to coat the mesh, in which thermal-responsive polymer PNIPAAm exhibits an extended hydrophilic chain conformation below its LCST in aqueous solution, and undergoes a phase transition to an insoluble and hydrophobic aggregate above its LCST in air or under water.[73] The hydrophilic poly(methyl methacrylate) (PMMA) with oleophilic property under water acts as the crosslink of PNIPAAm due its high glass transition temperature (T_g). The reversible switching wetting states of BCP of PMMA-b-PNIPAAm between hydrophilicity/oleophobicity and hydrophobicity/oleophilicity with change of temperature is shown in Fig. 1.32a. Considering that PMMA with high T_g is less sensitive to temperature, PMMA is supposed to have stable oleophilicity and remain unchanged with thermal expansion. Repeatedly cycling at different temperatures shows that the as-prepared BCP of PMMA-b-PNIPAAm film undergoes a good reversible wetting transformation between hydrophilicity/oleophobicity and hydrophobicity/oleophilicity (Fig. 1.32b), which indicates excellent reproducibility and stability. The mechanism explanation is shown in Fig. 1.32c. Below the LCST, as seen in Fig. 1.32c, intermolecular hydrogen bonds with water lead to a hydrated, swollen state of PNIPAAm, which provides the surface with a high ratio of water content. A water-rich surface will effectively block the access of the oil. Consequently, the surface exhibits hydrophilic and oleophobic properties. Above the LCST, intramolecular hydrogen bonds lead to a dehydrated, collapsed state of PNIPAAm. Thus, PNIPAAm is hydrophobic, while PMMA is hydrophilic. PNIPAAm shrinks as the temperature increases, but it still possesses a larger surface area. Therefore, the BCP of PMMA-b-PNIPAAm film is hydrophobic. PMMA shows underwater oleophilic property. PNIPAAm with an incomplete degree of dehydration shows oleophobicity.[25] In water, the BCP of PMMA-b-PNIPAAm self-assembles into a lamellar structure. PNIPAAm domains lay in between the hard walls of PMMA on a nanometer scale. Water tends to release from the confined space. The degree of dehydration of PNIPAAm can become quite high with the aid of PMMA in the nanostructure. The nanoPNIPAAm domains become more oleophilic. Therefore, the BCP of PMMA-b-PNIPAAm film is oleophilic. The temperature-responsive wettability is related to not only to the conformation transition of PNIPAAm and the apparent PNIPAAm domain area changes, but also to the cooperation between PNIPAAm and PMMA.[74]

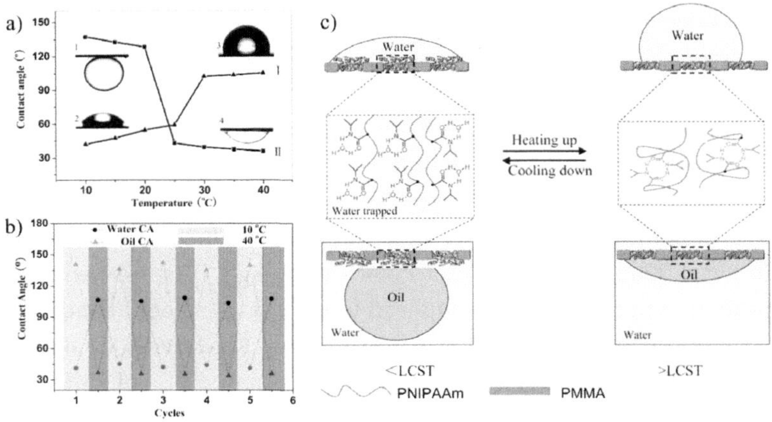

Figure 1.32 (a) Temperature dependences of water and oil CAs for a BCP of PMMA-b-PNIPAAm film. (b) Reversible water and oil CA transition of the BCP of PMMA-b-PNIPAAm film at different temperatures, indicating excellent reproducibility and stability. (c) Diagram of reversible formation of intermolecular hydrogen bonding between PNIPAAm chains and water molecules below the LCST, which leads to hydrophilicity/oleophobicity, and intramolecular hydrogen bonding between C=O and N–H groups in PNIPAAm chains above the LCST, which leads to hydrophobicity/oleophilicity.

This on–off switch shows excellent temperature controllable oil/water separation, as shown in Fig. 1.33. The bottleneck of a glass tube was sealed with a piece of BCP of PMMA-b-PNIPAAm-coated mesh and then put into the mixture of water and oil (dyed with iodine). When temperature is below LCST, water penetrates through the mesh and is collected in the bottle, while the oil will be collected when temperature is higher than LCST. In addition, the BCP of PMMA-b-PNIPAAm-coated mesh gives good separation ability after 10 times oil/water separation. On the other hand, this kind of temperature controllable dual oil/water separation system can be easily fabricated by direct casting of the BCP of PMMA-b-PNIPAAm on industrial steel mesh. This work offers promising applications in oil/water separation, micro-/nanofluidic devices.

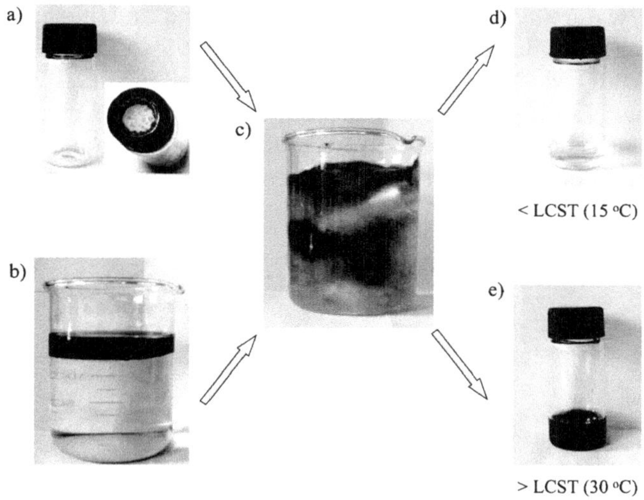

Figure 1.33 (a) A glass bottle was sealed with a piece of BCP of PMMA-b-PNIPAAm-coated mesh. (b & c) Mixture of water and oil (dyed with with iodine) before and after stirring. (d) Water was collected below the LCST. (e) Oil was collected above the LCST.

Tian and co-workers[35] demonstrated photo-induced "water-removing" type of oil/water separation based on underwater superoleophobic surface based on aligned ZnO nanorod array-coated stainless steel mesh. The fabrication process is described in the Section 1.2.2. This kind of smart oil/water separation materials exhibits switchable superhydrophobicity–superhydrophilicity, and underwater superoleophobicity at the special oil–water–solid three-phase interface. Figure 1.34 demonstrates the photo-induced oil/water separation process with a home-made device. The aligned ZnO nanorod array coated mesh film was fixed with a glass funnel. Firstly, a little amount of water was poured into the upper glass funnel to form a water film on the coated mesh film. Then the mixture of oil and water with ratio of 3:7 in volume was poured into the upper glass funnel. Due to the presence of the superhydrophobicity-induced strong negative capillary effect in the coated mesh film stored in the dark, the mixture of oil and water is well kept in the upper glass funnel. When UV light was applied to irradiate the coated mesh film, water selectively passed through the mesh and flowed into the lower bottle, whereas, the oil remained in the upper glass funnel. Finally, the oil/water mixture was successfully separated, and the separation efficiency calculated reached up as high as 95%. This promising photo-induced "water-removing" type of oil/water separation materials can be used as a microreaction system and controllable filtration, especially for real-time removal of a large amount of water from an in situ chemical reaction system for a high product of target material.

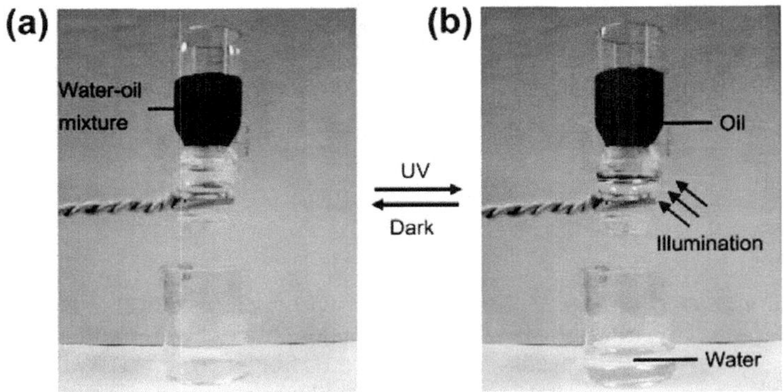

Figure 1.34 Demonstration of photo-induced oil/water separation device. (a) A little amount of water was poured into the upper glass funnel to form a water film on the ZnO nanorod array coated-mesh film. Then the mixture of crude oil and water was poured into the upper glass funnel. (b) When UV light irradiated the coated mesh film, water selectively passed through the mesh film, whereas the crude oil remained in the upper glass funnel.

1.3.3 Anti-Fouling Microfluidic Channel

The underwater superoleophobic surface can be also used in the functional microfulidic device. Wu and co-workers[30] developed a facile method to construct microchannel based on underwater superoleophobic surface through oxygen plasma treating PDMS. The PDMS is widely used in microfulidic devices due their economic, stable, and biocompatible properties.[75] As well known, PDMS is a typical intrinsic hydrophobic polymer with weak underwater oleophobic ability, seriously limiting its anti-oil applications. Wu et al. applied the oxygen plasma treatment to construct gecko-foot like arrays, and the underwater superoleophobic surface can be easily achieved as shown in Fig. 1.14. The resulting hierarchical surfaces shows OCA as high as 170°, a sliding angle less than 1°, and low adhesion force of about 1 μN. This kind of effective self-cleaning anti-oil surface is made into a demonstration of microfluidic channel for anti-oil contamination application as shown in Fig. 1.35. At the flat region, the oil cannot be washed away by water as the oil firmly adhered to the channel, while no oil existed on the microstructure region.

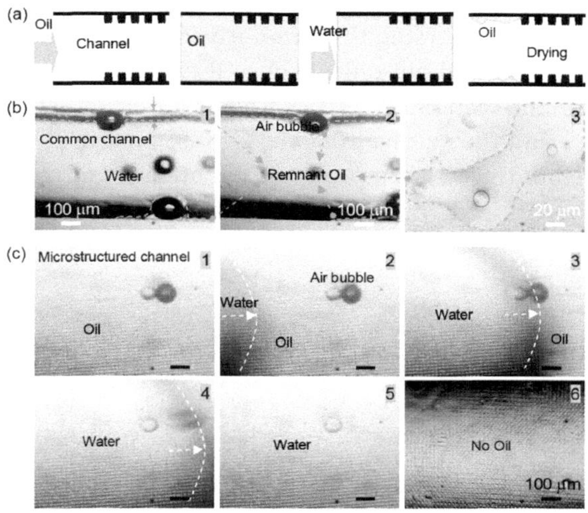

Figure 1.35 Comparison of a flat channel and microstructured channels. (a) Schematic illustration of the experimental process. Firstly, the oil was imported into the channel. Then, the water was used to clean the channel. Finally, the flat region and microstructured region were dried. (b) The oil droplet on the flat microfluidic channel. The oil firmly adhered to the channel and could not be washed out by water. (c) The self-cleaning anti-oil ability of the microfluidic channel with hierarchical microstructures. We could find that the oil on the flat region could not be removed while there is no oil on the microstructured channel. The air bubble was put into the microfluidic channels to better observe the self-cleaning oil function.

1.3.4 Underwater Oil Capture

Recently, Jin and co-workers[69] developed a novel organosilane surface which is superamphiphobic in air and superoleophilic under water. This surface mimics the desert beetle underwater and shows promising application in capturing oil droplets in water. Compared with the traditional oleophilic materials, this bio-inspired superoleophilic surface can repeatably capture oil droplets under water. The 1H,1H,2H,2H–perfluorodecyltrichlorosilane (FTS) was grafted on a substrate through hydrolysis and polycondensation to achieve a 3D network architectured surface. The FTS-derived surface shows superamphiphobic property with CA of water of 168.2 ± 1.3°, CA of oil of 148.1 ± 2.1°, respectively, as shown in Fig. 1.36. When immersing the FTS-derived surface into the water, the oil droplet spread out quickly once it contacted the surface. As well known, the wettability of a surface is governed by the surface free-energy and the surface roughness. SEM images show that the FTS-derived surface is composed of nanofibers and microbumps, which crisscross each other and form a 3D network architecture.

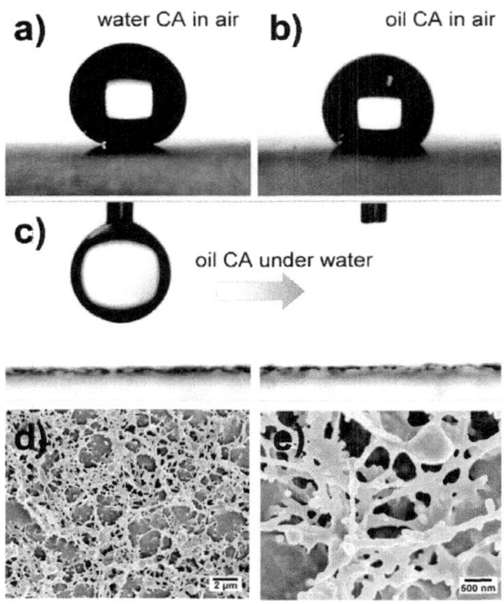

Figure 1.36 (a) Contact angles of water and (b) an oil droplet on the FTS-derived surface are 168.2 ± 1.3°, and 148.1 ± 2.1°, respectively. (c) While under water, the oil droplet spread out when contact with the surface, indicating the surface is superoleophilic under water. (d & e) SEM images of the FTS-derived surface indicate the surface is composed of nanofibers and microbumps, which crisscross each other and form an 3D network architecture.

An oil capture system of FTS-derived glass tube capturing oil droplets is demonstrated in Fig. 1.37. A layer of oil droplets (1,2-dichloroethane dyed yellow) was sprayed under water in a quartz container (step 1). Then, FTS-derived glass tube is immersed into water to touch and collect oil droplets (step 2). When moving the glass tube, the oil droplets gathered around the bottom of the FTS-derived glass tube and coalesced together (step 3). Finally, the oil droplets were easily captured by the FTS-derived glass tube (steps 4–6).

Figure 1.37 A demonstration process of the oil capture system of FTS-derived glass tube capturing oil droplets under water. Process 1: A layer of oil droplets is sprayed in bottom of a water container. Process 2: FTS-derived glass tube touches and captures the oil droplets underwater. Process 3: As the FTS-derived glass tube moves, oil drops are gathered together. Process 4–6: The oil droplets are collected from the water.

1.4 Conclusions and Perspective

In this chapter, the recent research progress and applications of bio-inspired underwater superoleophobic materials have been briefly summarized. Since the first report of underwater superoleophobic materials in 2009, many different strategies have been proposed to construct underwater superoleophobic materials. A variety of applications in the fields including oil/water separation, anti-fouling, underwater self-cleaning, underwater oil capture, and other fields, have been successfully demonstrated. Despite some progress in these fields has been achieved, the research of underwater superoleophobic materials is still in its infancy, and further fundamental research are needed to understand the wetting behavior of oil droplets underwater.

Like the research process of the superhydrophobic materials, the further investigation should be moved to the smart underwater superoleophobic surface, due to their unique properties and promising applications. The smart underwater superoleophobic materials can be designed and constructed with different stimuli-responsive, such as, pH, thermal, UV, electric, magnetic, and integration of several stimuli together. These intelligent surfaces show great potential applications in lab-on-chip systems, bio-sensors, nanodevices, etc.

References

1. de Gennes PG (1992) Soft matter, *Rev Mod Phys*, **64**(3):645–648.
2. Lim HS, Park SH, Koo SH, et al. (2010) Superamphiphilic Janus fabric, *Langmuir*, **26**(24), 19159–19162.
3. Liu B, Wei W, Qu X, Yang Z (2008) Janus colloids formed by biphasic grafting at a pickering emulsion interface, *Angew Chem Int Ed*, **47**(21), 3973–3975.
4. Liang F, Shen K, Qu X, et al. (2011) Inorganic Janus nanosheets, *Angew Chem Int Ed*, **50**(10):2379–2382.
5. Cheng Q, Li M, Zheng Y, Su B, Wang S, Jiang L (2011) Janus interface materials: Superhydrophobic air/solid interface and superoleophobic water/solid interface inspired by a lotus leaf, *Soft Matter*, **7**(13), 5948–5951.
6. Sun TL, Feng L, Gao XF, Jiang L (2005) Bioinspired surfaces with special wettability, *Acc Chem Res*, **38**(8), 644–652.
7. Barthlott W, Neinhuis C (1997) Purity of the sacred lotus, or escape from contamination in biological surfaces, *Planta*, **202**(1), 1–8.
8. Feng L, Li S, Li Y, et al. (2002) Super-hydrophobic surfaces: From natural to artificial, *Adv Mater*, **14**(24), 1857–1860.
9. Zhang JH, Wang JM, Zhao Y, et al (2008) How does the leaf margin make the lotus surface dry as the lotus leaf floats on water, *Soft Matter*, **4**(11), 2232–2237.
10. Erbil HY, Demirel AL, Avcı Y, Mert O (2003) Transformation of a simple plastic into a superhydrophobic surface, *Science*, **299**(5611), 1377–1380.

11. Xie Q, Xu J, Feng L, et al. (2004) Facile creation of a super-amphiphobic coating surface with bionic microstructure, *Adv Mater*, **16**(4), 302–305.

12. Teshima K, Sugimura H, Inoue Y, Takai O, Takano A (2003) Ultra-water-repellent poly(ethylene terephthalate) substrates, *Langmuir*, **19**(25), 10624–10627.

13. Shirtcliffe NJ, McHale G, Newton MI, Perry CC (2003) Intrinsically superhydrophobic organosilica sol–gel foams, *Langmuir*, **19**(14), 5626–5631.

14. Jiang L, Zhao Y, Zhai J (2004) A lotus-leaf-like superhydrophobic surface: A porous microsphere/nanofiber composite film prepared by electrohydrodynamics, *Angew Chem Int Ed*, **43**(33), 4338–4341.

15. Zhang X, Shi F, Yu X, et al. (2004) Polyelectrolyte multilayer as matrix for electrochemical deposition of gold clusters, toward super-hydrophobic surface, *J Am Chem Soc*, **126**(10), 3064–3065.

16. Liu MJ, Zheng YM, Zhai J, Jiang L (2010) Bioinspired super-antiwetting interfaces with special liquid-solid adhesion, *Acc Chem Res*, **43**(3), 368–377.

17. Xia F, Jiang L (2008) Bio-inspired, smart, multiscale interfacial materials, *Adv Mater*, **20**(15), 2842–2858.

18. Liu K, Yao X, Jiang L (2010) Recent developments in bio-inspired special wettability, *Chem Soc Rev*, **39**(8), 3240–3255.

19. Osborn JM, Schneider EL (1988) Morphological-studies of the Nymphaeaceae Sensu Lato. 16. The floral biology of Brasenia-Schreberi. *Ann Mo Bot Gard*, **75**(3), 778–794.

20. Liu MJ, Wang ST, Wei ZX, Song YL, Jiang L (2009) Bioinspired design of a superoleophobic and low adhesive water/solid interface, *Adv Mater*, **21**(6), 665–669.

21. Qunfeng C: *Unpublished Data*.

22. Jung YC, Bhushan B (2009) Wetting behavior of water and oil droplets in three-phase interfaces for hydrophobicity/philicity and oleophobicity/philicity†, *Langmuir*, **25**(24), 14165–14173.

23. Jin M, Wang J, Yao X, Liao M, Zhao Y, Jiang L (2011) Underwater oil capture by a three-dimensional network architectured organosilane surface, *Adv Mater*, **23**(25), 2861–2864.

24. Lin L, Liu M, Chen L, et al. (2010) Bio-inspired hierarchical macromolecule–nanoclay hydrogels for robust underwater superoleophobicity, *Adv Mater* **22**(43), 4826–4830.

25. Chen L, Liu MJ, Lin L, et al. (2010) Thermal-responsive hydrogel surface: Tunable wettability and adhesion to oil at the water/solid interface, *Soft Matter*, **6**(12), 2708–2712.
26. Xu L-P, Zhao J, Su B, et al. (2013) An ion-induced low-oil-adhesion organic/inorganic hybrid film for stable superoleophobicity in seawater, *Adv Mater*, **25**(4), 606–611.
27. Wang J, Lin L, Cheng Q, Jiang L (2012) A strong bio-inspired layered PNIPAM-clay nanocomposite hydrogel, *Angew Chem Int Ed*, **51**(19), 4676–4680.
28. Liu X, Zhou J, Xue Z, et al. (2012) Clam's shell inspired high-energy inorganic coatings with underwater low adhesive superoleophobicity, *Adv Mater*, **24**(25), 3401–3405.
29. Jin M, Feng X, Xi J, et al. (2005) Super-hydrophobic PDMS surface with ultra-low adhesive force, *Macromol Rapid Comm*, **26**(22), 1805–1809.
30. Wu D, Wu S-z, Chen Q-D, et al. (2011) Facile creation of hierarchical PDMS microstructures with extreme underwater superoleophobicity for anti-oil application in microfluidic channels, *Lab Chip*, **11**(22), 3873–3879.
31. Zhou J, Ellis AV, Voelcker NH (2010) Recent developments in PDMS surface modification for microfluidic devices, *Electrophoresis*, **31**(1), 2–16.
32. Liu MJ, Nie FQ, Wei ZX, Song YL, Jiang L (2010) In situ electrochemical switching of wetting state of oil droplet on conducting polymer films, *Langmuir*, **26**(6), 3993–3997.
33. Chen L, Liu M, Bai H, et al. (2009) Antiplatelet and thermally responsive poly(n-isopropylacrylamide) surface with nanoscale topography, *J Am Chem Soc*, **131**(30), 10467–10472.
34. Xue Z, Wang S, Lin L, et al. (2011) A novel superhydrophilic and underwater superoleophobic hydrogel-coated mesh for oil/water separation, *Adv Mater*, **23**(37), 4270–4273.
35. Tian D, Zhang X, Tian Y, et al. (2012) Photo-induced water-oil separation based on switchable superhydrophobicity-superhydrophilicity and underwater superoleophobicity of the aligned ZnO nanorod array-coated mesh films, *J Mater Chem*, **22**(37), 19652–19657.
36. Hejazi V, Nosonovsky M (2011) Wetting transitions in two-, three-, and four-phase systems, *Langmuir*, **28**(4), 2173–2180.
37. Hejazi V, Nyong AE, Rohatgi PK, Nosonovsky M (2012) Wetting transitions in underwater oleophobic surface of brass, *Adv Mater*, **24**(44), 5963–5966.

38. Xue ZX, Liu MJ, Jiang L (2012) Recent developments in polymeric superoleophobic surfaces, *J Polym Sci Pol Phys,* **50**(17), 1209–1224.

39. Cheng Q, Li M, Yang F, et al. (2012) An underwater pH-responsive superoleophobic surface with reversibly switchable oil-adhesion, *Soft Matter,* **8**(25), 2740–2743.

40. Liu M, Xue Z, Liu H, Jiang L (2012) Surface wetting in liquid–liquid–solid triphase systems: Solid-phase-independent transition at the liquid–liquid interface by Lewis acid–base interactions, *Angew Chem Int Ed,* **51**(33), 8348–8351.

41. Liu M, Liu X, Ding C, et al. (2011) Reversible underwater switching between superoleophobicity and superoleophilicity on conducting polymer nanotube arrays, *Soft Matter,* **7**(9), 4163–4165.

42. Ding C, Zhu Y, Liu M, Feng L, Wan M, Jiang L (2012) PANI nanowire film with underwater superoleophobicity and potential-modulated tunable adhesion for no loss oil droplet transport, *Soft Matter,* **8**(35), 9064–9068.

43. Gil ES, Hudson SM (2004) Stimuli-reponsive polymers and their bioconjugates, *Prog Polym Sci,* **29**(12), 1173–1222.

44. Xia F, Feng L, Wang ST, et al. (2006) Dual-responsive surfaces that switch superhydrophilicity and superhydrophobicity, *Adv Mater,* **18**(4), 432–436.

45. Hou X, Liu Y, Dong H, Yang F, Li L, Jiang L (2010) A pH-gating ionic transport nanodevice: Asymmetric chemical modification of single nanochannels, *Adv Mater,* **22**(22), 2440–2443.

46. Aizawa H, Kurosawa S, Kobayashi K, et al. (2000) Turning of contact angle on glass plates coated with plasma-polymerized styrene, allylamine and acrylic acid, *Mater Sci Eng C,* **12**(1–2), 49–54.

47. Lee JW, Kim SY, Kim SS, Lee YM, Lee KH, Kim SJ (1999) Synthesis and characteristics of interpenetrating polymer network hydrogel composed of chitosan and poly(acrylic acid), *J Appl Polym Sci,* **73**(1), 113–120.

48. Philippova OE, Hourdet D, Audebert R, Khokhlov AR (1997) pH-responsive gels of hydrophobically modified poly(acrylic acid), *Macromolecules,* **30**(26), 8278–8285.

49. Bajpai AK (1997) Interface behaviour of ionic polymers, *Prog Polym Sci,* **22**(3), 523–564.

50. Fowkes FM (1963) Additivity of intermolecular forces at interfaces. I. Determination of the contribution to surface and interfacial tensions of dispersion forces in various liquids, *J Phys Chem,* **67**(12), 2538–2541.

51. Huang X, Lim T-T (2006) Performance and mechanism of a hydrophobic–oleophilic kapok filter for oil/water separation, *Desalination*, **190**(1–3), 295–307.

52. Toyoda M, Aizawa J, Inagaki M (1998) Sorption and recovery of heavy oil by using exfoliated graphite, *Desalination*, **115**(2), 199–201.

53. Inagaki M, Kawahara A, Nishi Y, Iwashita N (2002) Heavy oil sorption and recovery by using carbon fiber felts, *Carbon*, **40**(9), 1487–1492.

54. Gui X, Wei J, Wang K, et al. (2010) Carbon nanotube sponges, *Adv Mater*, **22**(5), 617–621.

55. Lee C, Baik S (2010) Vertically-aligned carbon nano-tube membrane filters with superhydrophobicity and superoleophilicity, *Carbon*, **48**(8), 2192–2197.

56. Zhao Y, Hu C, Hu Y, Cheng H, Shi G, Qu L (2012) A versatile, ultralight, nitrogen-doped graphene framework, *Angew Chem Int Ed*, **51**(45), 11371–11375.

57. Reynolds JG, Coronado PR, Hrubesh LW (2001) Hydrophobic aerogels for oil-spill clean up—synthesis and characterization, *J Non-Cryst Solids*, **292**(1–3), 127–137.

58. Venkateswara Rao A, Hegde ND, Hirashima H (2007) Absorption and desorption of organic liquids in elastic superhydrophobic silica aerogels, *J Colloid Interf Sci*, **305**(1), 124–132.

59. Feng L, Zhang Z, Mai Z, et al. (2004) A super-hydrophobic and super-oleophilic coating mesh film for the separation of oil and water, *Angew Chem Int Ed*, **43**(15), 2012–2014.

60. Yuan J, Liu X, Akbulut O, et al. (2008) Superwetting nanowire membranes for selective absorption, *Nat Nanotechnol*, **3**(6), 332–336.

61. Zhang Y, Wei S, Liu F, et al. (2009) Superhydrophobic nanoporous polymers as efficient adsorbents for organic compounds, *Nano Today*, **4**(2), 135–142.

62. Ono T, Sugimoto T, Shinkai S, Sada K (2007) Lipophilic polyelectrolyte gels as super-absorbent polymers for nonpolar organic solvents, *Nat Mater*, **6**(6), 429–433.

63. Sonmez HB, Wudl F (2005) Cross-linked poly(orthocarbonate)s as organic solvent sorbents, *Macromolecules*, **38**(5), 1623–1626.

64. Hayase G, Kanamori K, Fukuchi M, Kaji H, Nakanishi K (2013) Facile synthesis of marshmallow-like macroporous gels usable under harsh conditions for the separation of oil and water, *Angew Chem Int Ed*, **52**(7), 1986–1989.

65. Zhang X, Li Z, Liu K, Jiang L (2013) Bioinspired multifunctional foam with self-cleaning and oil/water separation, *Adv Funct Mater,* **23**(22), 2881–2886.

66. Wei QF, Mather RR, Fotheringham AF, Yang RD (2003) Evaluation of nonwoven polypropylene oil sorbents in marine oil-spill recovery, *Mar Pollut Bull,* **46**(6), 780–783.

67. Kulawardana EU, Neckers DC (2010) Photoresponsive oil sorbers, *J Polym Sci A Polym Chem,* **48**(1), 55–62.

68. Wen Q, Di J, Jiang L, Yu J, Xu R (2013) Zeolite-coated mesh film for efficient oil-water separation, *Chem Sci,* **4**(2), 591–595.

69. Jin M, Li S, Wang J, Xue Z, Liao M, Wang S (2012) Underwater superoleophilicity to superoleophobicity: Role of trapped air, *Chem Commun,* **48**(96), 11745–11747.

70. Zhang L, Zhang Z, Wang P (2012) Smart surfaces with switchable superoleophilicity and superoleophobicity in aqueous media: Toward controllable oil/water separation, *NPG Asia Mater,* **4**:e8.

71. Cheng M, Gao Y, Guo X, Shi Z, Chen J-f, Shi F (2011) A functionally integrated device for effective and facile oil spill cleanup, *Langmuir,* **27**(12), 7371–7375.

72. Xue B, Gao L, Hou Y, Liu Z, Jiang L (2013) Temperature controlled water/oil wettability of a surface fabricated by a block copolymer: Application as a dual water/oil on–off switch, *Adv Mater,* **25**(2), 273–277.

73. Sun T, Wang G, Feng L, et al. (2004) Reversible switching between superhydrophilicity and superhydrophobicity, *Angew Chem Int Ed,* **43**(3), 357–360.

74. Sun T, Song W, Jiang L (2005) Control over the responsive wettability of poly(*N*-isopropylacrylamide) film in a large extent by introducing an irresponsive molecule, *Chem Commun,* (13), 1723–1725.

75. McDonald JC, Whitesides GM (2002) Poly (dimethylsiloxane) as a material for fabricating microfluidic devices, *Acc Chem Res,* **35**(7), 491–499.

Chapter 2

Butterfly Wing Effect: Anisotropically Oriented Micro- and Nanostructures

Butterfly is a beautiful insect in nature.[1] Interestingly, many smart multifunctions may be found on wings,[2] for example, water repellency. These biological functions are result of micro- and nanostructure, which serve well for survival of butterfly in nature.[3–6] In recent year, *Morpho* blue butterfly wing (Fig. 2.1, top) has been revealed into special wettability. For example, butterfly wing has self-cleaning property similar to lotus leaf. In addition, butterfly wing has anisotropic multistructure that leads to anisotropic wetting just as observed on plant leaves,[7] flower petals,[8,9] and bird feathers.[10,11] More interestingly, butterfly wing displays the unidirectionality of droplet shedding-off, which can be defined as anisotropic wettability. For example, butterfly wings of iridescent blue *Morpho aega* butterflies have a typical character of excellent directional adhesion for fluidic controlling.[3,12] This directionality on wettability can be attributed to the multilevel regular micro-/nanostructures of wings. So, such multiple-anisotropic wettability can be herein defined as "effect of butterfly wing," which guards us to design novel functional surfaces. For extending aspect, some structures similar to wing of butterfly can be extracted; for example, oriented structure, geometry, ratchet, step structure, and so on. It has become an interest area how to design the structured surfaces via some novel methods.

Bio-Inspired Wettability Surfaces: Developments in Micro- and Nanostructures
Yongmei Zheng

ISBN 978-981-4463-60-7 (Hardcover), 978-981-4463-61-4 (eBook)
www.panstanford.com

Just as we study, all these natural structured surfaces derive the directional properties of wettability due to asymmetric micro- or nanostructures such as ratchets or hairy forests. Anisotropism is one of the most important properties on these surfaces with oriented structures, and results in special wetting and dewetting characteristics in different directions, which has drawn significant interest in a broad range of applications, such as icephobic/anti-icing surfaces, light- and electricity-modulated liquid transport, selective particle transport, anti-fogging/self-cleaning at low temperatures, reversible rolling of a drop, and high temperature–induced or vibration-induced movement of a drop, and so on (Fig. 2.1). The anisotropic dewetting property may also bring interesting insights into design of lossless liquid transportation channels and novel microfluidic valves.

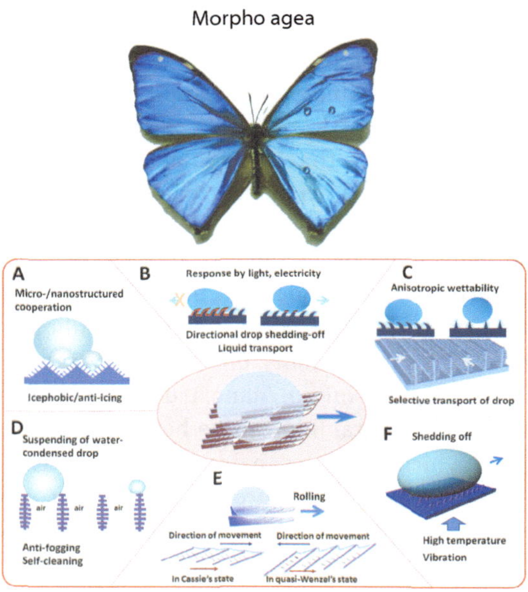

Figure 2.1 Top: Photo of *Morpho* butterfly. Bottom: Schematic of the design of a multilevel structured surface for functional wettability.

In this chapter, we introduce the wetting properties of asymmetric ratchet structures surrounding butterfly wing to elucidate the directional adhesion, directional water repellency, low-temperature superhydrophobicity, and also surrounding ryegrass leaf with taper-ratchet array to explain the mechanism of directional shedding-off of water. We also introduce some methods of how to obtain the asymmetric structured surface and some smart functions such as fluid controlling, droplet transport, anti-icing, and anti-fogging. All these have inspired us to design novel functional materials.

2.1 Butterfly Wing

Although color of butterfly wing is induced by structure, wettability of butterfly wing is of importance to its life activities. In 1996, Wagner et al.[2] reported the superhydrophobic (SH) and self-cleaning characteristics of butterfly wings. Due to large wingspan, butterflies cannot reach the whole wing surface with their legs, and therefore, cannot clean them by rubbing. Thus, for activities, they are completely dependent on the self-cleaning properties of the wing surface. Some interesting phenomenon can be observed in nature. In 2006, Zheng et al.[3] found that the wettability anisotropism phenomena of butterfly *Morpho aega*, which are attributed to the multilevel regular micro-/nanostructures on tiny scales of wings. Up to now, many researches pay attention into the anisotropic wettability.[11–16]

2.1.1 SH and Directional Adhesion

Butterfly wings are not only SH but also show directional adhesion. Zheng et al.[17] theoretically demonstrated that the elasticity deformation of the substrate can also be used to guide droplets motion. According to the principle of minimum potential energy, the interaction between droplet and substrate can drive the droplet to move directionally from the stiffer to the softer end of the beam. This study may provide another reasonable explanation to the effect of butterfly wing, that is, the elasticity deformation and pinning mechanism of the microstripes on the wing performs the directional motion of droplets.

It was reported that if a droplet is pinned on or rolls off, the SH surface is ascribed to both the distinct contact modes. Thus, the Wenzel state exhibits an extremely high adhesive property. In Cassie and Baxter mode, the droplet will easily roll off a surface with extremely low adhesion if a discontinuous three-phase contact line (TCL) can effectively form. The dramatic decrease of the surface contact area of a droplet with the microstructure can greatly enhance the hydrophobicity of surfaces, which results in superhydrophobicity of butterfly wing with an apparent contact angle above 150°. It is observed that a water droplet can easily roll off the surface of the wings along the radial-outward (RO) direction of wings, but is tightly pinned in the opposite direction. The experiment can be carried out by tilting the surface of wing, in addition, by gas flowing into droplet on surface of wing (Fig. 2.2). It is demonstrated that this unique ability is due to the direction-dependent arrangement of flexible nanotips on the top of ridging nanostrips and microscales overlapped on the wings at the one-dimensional (1D) level. When the wing is tilted downward, the microscales with ridged nanostripes are spatially separated from each other and the oriented nanotips tend to be unwound with flexible microscales. In this case, air can be efficiently trapped in these nanoscale voids among the nanotips extended by lamellae and the ridged nanostripes. This ensures the superhydrophobicity of the wings, with a high contact angle above 150°. The formation of the extremely discontinuous TCL makes the droplet easily roll off the wings along the RO direction. However, when the wing is tilted upward, the flexible nanotips and microscales take on a close arrangement and the nanotips on the top of the nanostripes are raised with the flexible microscales to closely contact the droplet. The complete "wet" contact of water with the nanotips and lamellae would increase the fraction of solid/liquid surface area. As a result, a quasicontinuous TCL is formed. Thus, the droplet is pinned on the wing.

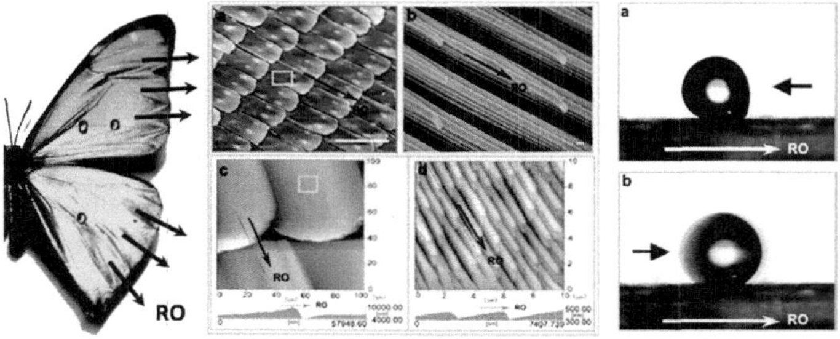

Figure 2.2 Blue *Morpho agea* butterfly wing (the left) has multilevel oriented structures (the middle). The RO direction (indicated with arrows). A droplet rolls off along RO direction and is pinned against RO direction via the act of gas flowing (the right).[3]

2.1.2 Anisotropism of Microstructures on Butterfly Wing Surface

Morpho butterflies have attracted much interest in investigating the wetting mechanism that contributes to foundational and biomimetic researches. Especially, the anisotropic adhesion resulting from the oriented multilevel micro-/nanostructure of butterfly wing bringing the broad significance in designing the self-cleaning, microfluidic-controlled surface, other artificially multifunctional surfaces. Here, we introduce the anisotropic water repellency, which resulted from the flexible oriented micro- and nanostructure of *Morpho* butterflies' wings. We discusses how difference of anisotropic water repellency is induced by slight regulation of nanotips in addition to microstructure on three kinds of wings of *Morpho* butterflies (named *aega*, *sulkowskyi*, and *rhotcnor*), where the efficiently 1D asymmetric topologies can be tuned by the tilted degree, periods of nanotips, and gaps on stripes of wings. This property offers the insight into how to modulate the anisotropy in water repellency on wings of different butterflies. It helps us to understand the life activities of butterflies in humid environments. And also, it is helpful to fabricate the novel water-repellent, microfluid-controlled, liquid transport surfaces for broad applications.

The iridescent blue *Morpho aega*, *sulkowskyi*, and *rhotcnor* butterflies come from Brazil. The wings of butterflies were firstly observed by optical microscopy at low resolution. There were many venations distributed on these butterflies' wings. The opens of venations were along the RO direction of body's central axis. Static contact angle and droplet-roll-off property at every direction are observed on wings of butterflies (Fig. 2.3). Interestingly, when a microdroplet of 3 μL volume is placed on the surface of butterfly wing, a strongly anisotropic roll-off at 1D level is found along the RO direction of body's center axis (Fig. 2.3, denoted with the arrows on the wing butterfly, the right top). The result is when the wing is tilted along the RO direction, the droplet easily rolls off, at a tilted angle of ~10° on wing of *M. sulkowskyi* and ~5° on wing of *M. rhotcnor*, respectively, while ~9° on wing of *M. aega*. But against the RO direction, ~30° on wing of *M. sulkowskyi* and ~18° on wing of *M. rhotcnor*, respectively, while the pinned on wing of *M. aega*. A larger ranged anisotropy (from roll-off angle of 9° to the pinning) is on wing of *M. aega*. An intermediate range (roll-off angle of 10–30°) is on that of *M. sulkowskyi*. A smaller range (roll-off angle of 5–18°) is on that of *M. rhotcnor*. The right images are the typical processes of droplet rolling-off behaviors on wing of *M. aega*. Droplet easily rolls off at a small roll-off angle (the top) when wing is tilted along the RO direction. On the contrary, droplet is easily pinned on surface even at upright (the bottom) when wing is tilted against the RO direction.

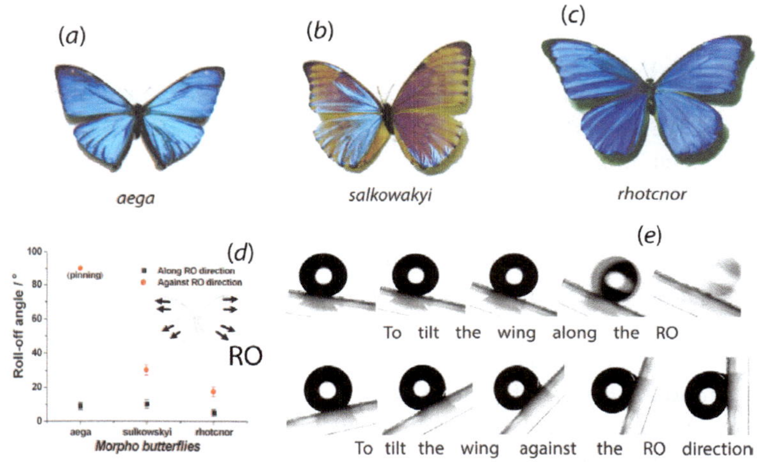

Figure 2.3 Anisotropic water repellency on *Morpho* butterflies' wings. (a) *M. aega* butterfly with SH wings. (b) *M. sulkowskyi* butterfly with SH wings. (c) *M. rhotcnor* butterfly with SH wings. (d) The graded anisotropic roll-off angles in 1D level along the contrasting RO direction on butterflies' wings, respectively (right). The RO direction is denoted with arrows on wing of butterfly (left top). (e) Serial captured photos that 1D anisotropic roll-off of droplet captured by optical contact angle meter when wing is tilted along/against RO direction.

Such differences of three-group anisotropies might be attributed to the finer structures on individual butterfly wing. Figure 2.4 shows the scanning electron microscopy (SEM) images of microstructure on wings of *M. aega*. The surface of wing is covered with overlapping micrometer scales (~150 μm in length and ~70 μm in width, in Fig. 2.4a), and there are thousands of separated stripes on scale (~184 nm in width, ~585 nm in clearance, in Fig. 2.4b) stacked with different-length lamella. The nano-tips from the end of lamella on stripes is tilted at an angle of ~5°, there is ~2.3 μm period of nanotips on stripes (Fig. 2.4c).

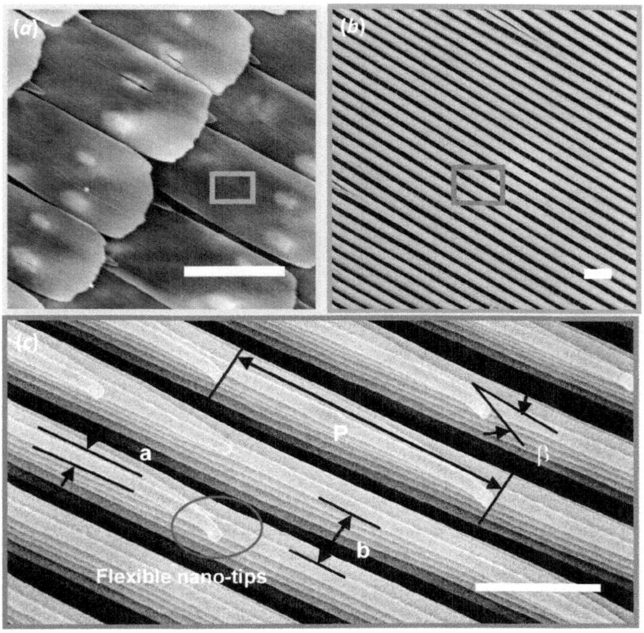

Figure 2.4 SEM images of finer microstructure on *M. aega* butterfly's wing. (a) Wing is covered with monolayer overlapping scales and each scale is about 150 μm × 70 μm in size. (b) Thousands of separated stripes on scale. (c) The stripe has the width a of ~109 nm, the clearance b of ~607 nm, and stacked with nanometer lamella and some uptilted tips to the lamella with a periodicity p of ~2.3 μm, tilted angle β of 5° on stripe to the base. Scale bar, (a) 150 μm, (b) 1 μm, (c) 1 μm.

Figure 2.5 is the SEM images of microstructures on *M. sulkowskyi* wings. Wings are covered with monolayer overlapping scales and each scale is about 152 μm × 84 μm in size (Fig. 2.5a), and there are thousands of large-separated stripes on scale, with ~86 nm in width, ~688 nm in clearance (Fig. 2.5b). The stripes are stacked with nanometer lamella and some flat tips (a tilted angle of ~2° to the lamella, and the nano-tips of ~2.4 μm periodicity along the stripe (Fig. 2.5c).

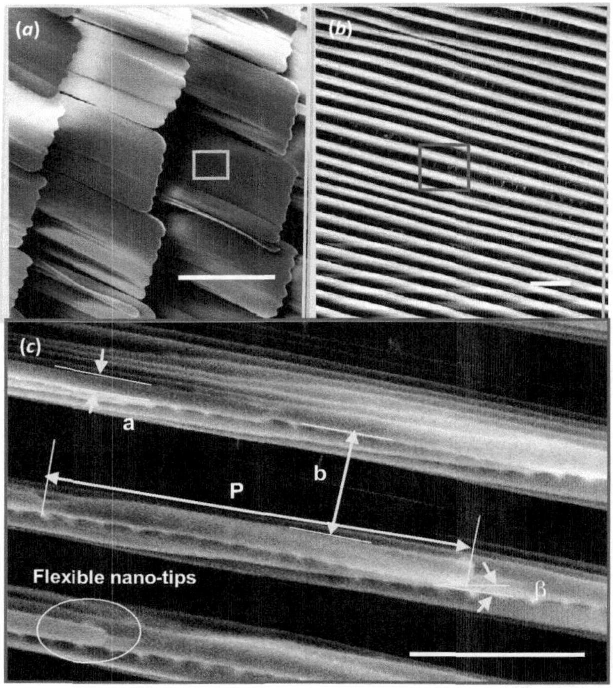

Figure 2.5 SEM images of finer microstructure on *M. sulkowskyi* butterfly's wing. (a) Wing is covered with monolayer overlapping scale and each scale is about 152 μm × 84 μm in size. (b) Thousands of largely separated stripes on scale. (c) The stripe has the width a of ~86 nm, the clearance b of ~688 nm, and stacked with nanometer lamella and some flat tips to the lamella with a periodicity p of ~2.4 μm, tilted angle β of 2° on stripe to the base. Scale bar: (a) 100 μm, (b) 1 μm, (c) 1 μm.

Figure 2.6 is the SEM images of microstructures on *M. rhotcnor* wings. Wings are covered with dual-layer overlapping scales: glass-scale on ground-scale. Glass-scale is about 177 μm × 49 μm in size (Fig. 2.6a). There are thousands of largely separated stripes on glass-scale, having ~115 nm in width, ~2.3 μm in clearance (Fig. 2.6b). The stripes are stacked with the highly tilted nanometer lamella, and the nanotips from the end of the lamella exhibits a distinct periodicity of ~1.2 μm and has tilted angle of ~10° (Fig. 2.6c). These present a micro- and nanostriped structure, accompanied with the oriented arrangements from the multi-layer lamella on stripes.

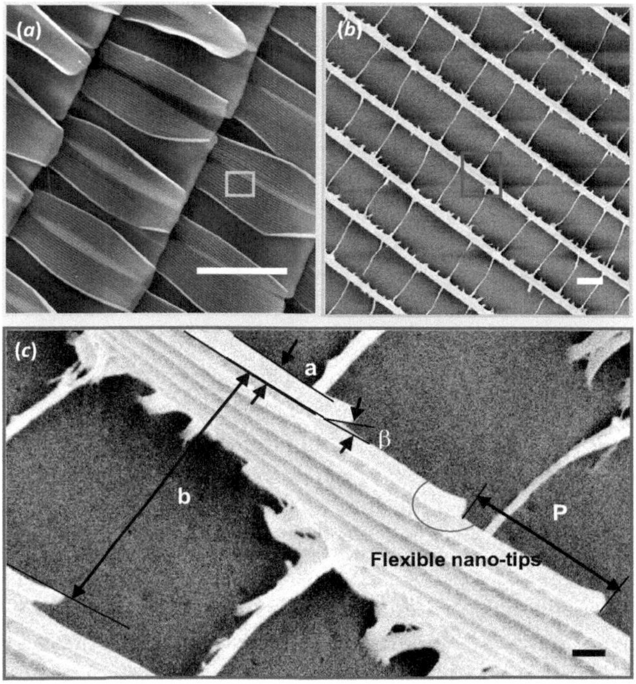

Figure 2.6 SEM images of finer microstructure on *M. rhotcnor* butterfly's wing. (a) Wing is covered with dual-layer overlapping scale: glass-scale on ground-scale and glass-scale is about 177 μm × 49 μm in size. (b) Thousands of largely separated stripes on glass scale. (c) The stripe has the width of ~115 nm, the clearance b of ~2.3 μm, and stacked with tilted nanometer lamella and distinctly oriented multilayer microstructure and flat-tips to the lamella with a periodicity *p* of ~1.2 μm, tilted angle β of 10° on stripe to the base. Scale bar: (a) 100 μm, (b) 1 μm, (c) 100 nm.

In viewpoint of wettability, a foundational theory is carried out via a combination of the surface chemistry and the surface roughness or topography. It is known that classical Young's equation expresses the contact angle θ_0 of liquid droplet, that is, $\cos\theta_0 = (\gamma_{SV} - \gamma_{SL})/\gamma_{LV}$, where γ_{SL}, γ_{SV}, and γ_{LV} are the interfacial free energies per unit of the solid/liquid, solid/gas, and liquid/gas interfaces, respectively. Such a flat surface often accompanies a lower contact angle than that of rough one when the surface-energy is low. The wettability can be modulated by the roughness of surface. Generally, there are two models: Wenzel's[18] and Cassie's.[19] Wenzel proposed that the contact angle θ_W on a rough surface can be modified from the Young's equation as $\cos\theta_W = r\cos\theta_0$, where r is the roughness factor, defined as the ratio of the actual area of a rough surface to the geometric projected area (in general, $r > 1$). This model indicates a state of wet contact in solid–liquid interface, accompanied with high adhesion. In contrast, Cassie and Baxter proposed a heterogeneous surface composed of solid and air phase, given by the equation: $\cos\theta_{CB} = \varphi\cos\theta_0 + (\varphi - 1)$, where φ is the solid fraction, $(1 - \varphi)$ is the air fraction on surface. On such a composite surface, droplet penetrates little the gaps of microstructure. Due to the air-pocket between the microstructure, a decreasing solid–liquid contact would induce a low adhesion action on microdroplet, thus droplet easily rolls off the surface. So, it indicates that the roughness r and fraction φ become the main structure parameters to control the variety of wettability on a surface,[18] but it is not excluded the influence of external force. Recent research has verified that water can penetrate the gaps of the microstructure by pressing the droplet on composite surface with a low surface-energy, forming a close contact, accordingly, produces the high adhesion at the interface of solid–liquid. In this case, the contact angle of water is intermediate (in metastable state for the droplet), but is not reversible when the contact angle reaches the critical value $\theta_c = (\varphi - 1)/(r - \varphi)$.[19]

The structure characteristics of *Morpho* butterflies' wings play an important role in nonwetting properties, making the static contact angles θ_{CB} of water about 152 ± 1.7° for *M. aega*, 155.2 ± 2.5° for *M. sulkowskyi* and 168.2 ± 3.4° for *M. rhotcnor*, respectively. Based on Cassie's law, the contact angle is regulated by

$$\cos\theta_{W-CB} = g(\cos\theta_{CB} + 1) - 1 \tag{1}$$

$$\cos\theta_{CB} = \varphi(\cos\theta_0 + 1) - 1 \tag{2}$$

Where φ is roughly estimated by the width and clearance of stripe in structure parameters, that is, $\varphi \sim a/(a + b)$, equals to ~0.2, ~0.1, and ~0.05 for the *M. aega*, *sulkowskyi*, and *rhotcnor*, respectively. Given the water contact angle θ_0 of ~105° on flat waxy surface,[20] the contact angle θ_{CB} is calculated by Eq. (2), about 148.4°, 157.8°, and 164.4°, respectively. Given that g is given by the integrate expression under a consideration of varied tilted degree: $g = \int_0^{\beta} \sin\beta d\beta$, where g is the fraction on stripes, related to the tilted angle of nanotips, approximately defined as the ratio of contact length (given, near to the height of nanotip) on nanotips to per period between nanotips. The upper limit of integrating β is the maximum tilted angle of flexible nanotips. So, Eq. (2.1) expresses the contact angle $\theta_{W\text{-}CB}$ that becomes the variable with the tilted angle. The different tilted angles would have different contact angles on the flexible-oriented stripe composite surfaces, which are in good agreement with the testing contact angle value by the contact angle meter. It indicates that the width (a), the clearance (b) of stripes is main to determinate the wettability for individual butterfly.

An enlarging clearance between stripes may create a large-ratio trapped-air. The manner in terms of the example of butterfly wing surface would enhance the contact angle by more than 50° as shown in Fig. 2.7a. According to the Cassie's model, we can simply estimate the range of water contact angle dependence of the clearance between stripes on wing, and the relation between the parameters in Cassie's equation is shown in Fig. 2.7b. It is illuminated that the relation between the solid fraction (φ) and the contact angle (θ') of rough surface and contact angle (θ) of flat surface (estimated above, agreement with that given by Young's equation). Distinctly, for a low surface-energy surface, the decreasing solid fraction φ will induce an increasing contact angle (θ'). As for a given fraction (φ), the decrease of surface-energy in low–surface-energy region (arrow line in Fig. 2.7b), the contact angle has an increasing range of 10~30°. But a change in fraction φ would induce a large range of 90~180° contact angle. This analysis predicts a tunable property of wettability by the striped microstructure. As for the roll-off of droplet, generally it is related to the difference in forces per unit length at the two sides of droplet, proportional to γ_{LV} ($\cos\theta_L - \cos\theta_u$), where θ_L and θ_u are the contact angle at the lower and upper sides of droplet. When the upper angle reached the receding angle θ_R and the lower angle reaches the advancing angle θ_A, the droplet just begins to move, is determined by the contact angle hysteresis,[20] $\Delta\theta_H = \theta_A - \theta_R$. The hysteresis can be overcome by weight of droplet at special rough-structure surface. But the weight may be neglected when the diameter D of droplet is far less than the millimetric capillary length (κ^{-1}). In this case, the topography would greatly change the moving state of droplet.[21] The solid fraction or roughness from the topographies would regulate the types of TCL along the direction of droplet moving if microstructure is asymmetric or fluctuated or flexible dynamic. A continuous TCL will produce the high solid–liquid contact (easily induces the sticky for droplet), while the discontinuous TCL will produce the low solid–liquid contact (easily induces the slippery for droplet).[3] So, the different anisotropic behaviors of droplets roll-off on a well-designed microstructure surface would be introduced by such two type of TCL.

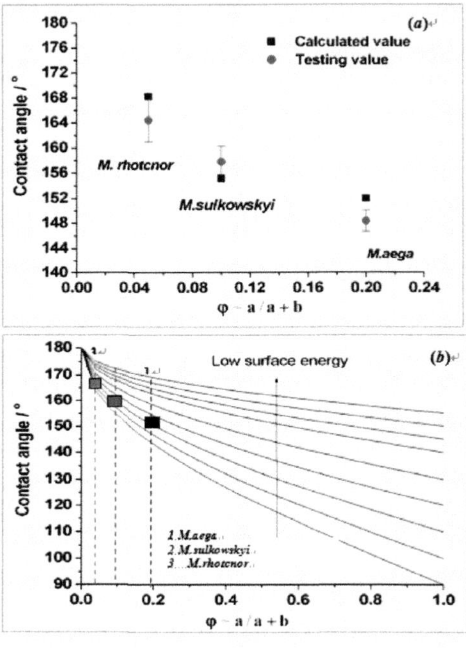

Figure 2.7 The contact angle θ' dependence of the solid fraction φ by Cassie's model. (a) Compare calculated and testing contact angle. The black solid square indicates the testing contact angle of 152.7 ± 1.7°, 155.2 ± 2.5°, and 168.2 ± 3.4°, and the red solid dot indicates the calculated contact angle of 148.4°, 157.8°, and 164.4° on wings of *M. aega, M. sulkowskyi,* and *M. rhotcnor* butterfly, respectively. (b) The series of reflexed curves represent the contact angle θ on flat surfaces (including the different surface-energies). Green, blue, and red dashed line is indicated with 1, 2, and 3, corresponding to the fraction φ and the green, blue, and red solid square indicates the contact angle on wings of *M. aega, M. sulkowskyi,* and *M. rhotcnor* butterfly, respectively.

Butterflies' wings have the hierarchical *oriented* micro- and nanostructure, seen by SEM observation. Microscales are overlapping along an orientation and cover over wings, and there are nanotips of oriented lamella on stripes. The asymmetric structure is distributed along the stripes of scale on wings, and this special structure strongly influences the roll-off property of droplet in our research.[3] This anisotropy of roll-off, dependent on the direction along/against the RO direction, proposed a generally potential mechanism of wettability on wing of butterfly. When the wing is tilted down, the oriented nanotips on the nanostripes and microscales separate from each other so that the water droplet deposit on the wing, and presents a composite contact mode with the top of the nanostripes and a "dry" contact with the air pocket trapped in the nanogrooves. Accordingly, an extremely discontinuous TCL forms between the droplet and the surface of wing. These factors as shown above all make the droplet easy to roll-off the surface. In contrast, when the wing is tilted upward, the flexible nanotips and microlevel scales of butterfly take a close arrangement so that the droplet presents the "wet" contact to the nanostripes and forms a quasicontinuous TCL, which pins it on the surface. While the intact area of the "dry" contact with the air pocket in the nanogroove remains, to ensure the high contact angle and superhydrophobicity on wings.

By comparison in details, we found that the micro- and nanostructure to water repellency on *M. aega* slightly distinguishes from those on *M. rhotcnor and M. sulkowskyi*, especially, in the nanotips state of asymmetrical topography on wings.

An essential structure diagram can be shown in Fig. 2.8a, among a, b is the width of stripes and the clearance between the stripes, p is period of nanotip on stripe, β is tilted degree of lamella on stripe to the base. It is revealed that a graded anisotropic roll-off is resulted from the oriented-tilted degree of nanometer tips, state of nanotips on stripe of the dissymmetrical topography. From the side-section diagram of stripe on scales, pinning droplet is due to a decreased tilted angle of nanotips and "wet" contact on stripes (Fig. 2.8b); in contrast, rolling droplet is due to an increased tilted angle of nanotips and "dry" contact on stripes (Fig. 2.8c). It is predicted that the uptilted tips of the lamella on *M. aega* might be more dominative to confine the direction of water-repellency than that on other *Morpho* butterflies such as *M. rhotcnor* and *M. sulkowskyi*. The scales and nanotips have a larger or smaller tilted degree, accordingly, the intermediated dependence of TCL is formed at 1D contrasting RO direction. Especially, as for *M. rhotcnor*, there is a larger gap of stripe on wings to weaken the solid fraction φ and a short period of tilted nanotip on stripe to decrease this difference of solid/liquid contact in two contrasting directions, and weaken the "dry" and "wet" contact on stripe. As a result, a small range in anisotropy of roll-off is produced. While as for *M. sulkowskyi*, there are flat nanotips and larger period of nanotips on stripe, is in intermediate anisotropy. These investigations provide a smart selection for anisotropy of water repellency by the regular fine micro- and nanostructure at 1D level.

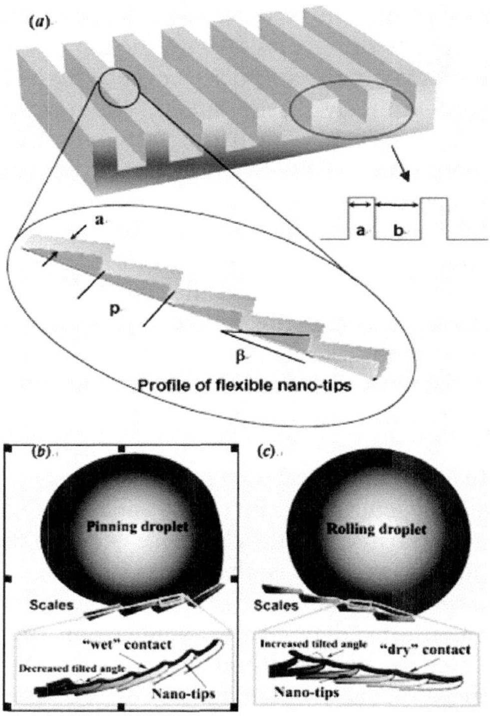

Figure 2.8 (a) The extracted microstructure on butterflies' wings including the striped structure on scale and tilted nanotips on stripes. Among a, b is the width of stripes and the clearance between the stripes, respectively, p is the period of nanotips on stripe, β is the lamella tilted degree on stripe to the base. (b) The rolling droplet due to an increased tilted angle of nanotips and "dry" contact on stripe. (c) The pinning droplet due to a decreased tilted angle of nano-tips and "wet" contact on stripe.

This investigation on wings of butterflies *M. aega*, *sulkowskyi*, and *rhotcnor* indicated that regular anisotropic water-repellency resulted from the oriented nano-tips on stripe with the different contact angle state at 1D level. Generally, the wings along the oriented direction can be low adhesion, but against the oriented direction can be high adhesion for directional water repellency. Especially, it is found that for stripe of a larger period and a small tilted degree of nanotips, there is a larger extent of anisotropic water-repellency along the direction of parallel to the stripe (corresponding to a low value of static contact angle). To the contrary, for stripe of a shorter period and a higher tilted degree of nanotips, it might induce a smaller extent of anisotropic water-repellency along the direction parallel to the stripe (corresponding to a high value of static contact angle). In addition, these also are regulated by the flexible degree of nanotips and the clearance between the stripes. Flexible nanotips and a moderate clearance might strengthen the anisotropy (corresponding to a moderate value of static contact angle). Whereas an introduction of short period of tilted nanotips would not enhance the anisotropy of water repellency although there is a high degree of tilted nanotips. It is concluded that a 1D strongly anisotropic water-repellent surface should be an optimal tilted or flexibly oriented nanotips, accompanied with a proper gap or clearance between stripes. Therefore, we deduce the biological meaning: the wing of *M. aega* butterfly needs a strong anisotropic water repellency to meet the safeguard of body and contribute to flight, but the wings of *sulkowskyi* and *rhotcnor* may reduce the anisotropy in water repellency to protect the body from the load of activities. These investigations help us to understand the life activities of butterflies in humid environments, and accept that this investigation is significant to design the surface of directional-microfluidic control in microfluidics.[3,12]

2.1.3 Low Temperature Water Repellency

For most low-surface energy surfaces, because the low temperature would induce the "water adsorption" on the solid surface, the superhydrophobicity would decrease and adhesion at the solid–liquid interfaces would increase.[22–24] Interestingly, micro-/nanostructure effects achieve water repellency of butterfly wing in special environment. For example, *Morpho nestira* butterfly wings with multilevel structures can lead the water shedding-off when wing is under low temperature and changeable relative humidity (Fig. 2.9). [10] The mechanism of the extended water repellency is elucidated based on the micro-/nanostructure effect that can be adapted to low temperature and relative humidity (Fig. 2.10). When a drop bridges ridges with dry nanotips at ambient temperature, the TCL is raised up to uplift the contact angle of water. But when a drop bridges the ridges with the wetted nanotips at low temperature, the TCL would be lowered down because of "water adsorption." In low relative humidity, the tiny condensed drops are allowed to coalesce to a suspended drop. This coalescence would finally sustain the trapped air to raise the contact angle of the water on the array of ridges; thus, the result is the role of Cassie's state may be prolonged effectively, then to weaken that of local quasi-Wenzel's state on nanotips. It is shown that the multilevel micro-/nanostructures on wing scales effectively retard low-temperature–induced wetting, and achieve water repellency, accompanied with low adhesion under environmental humidity conditions.

Figure 2.9 SEM shows blue *Morpho nestira* butterfly wings have multilevel micro-/nanostructure.

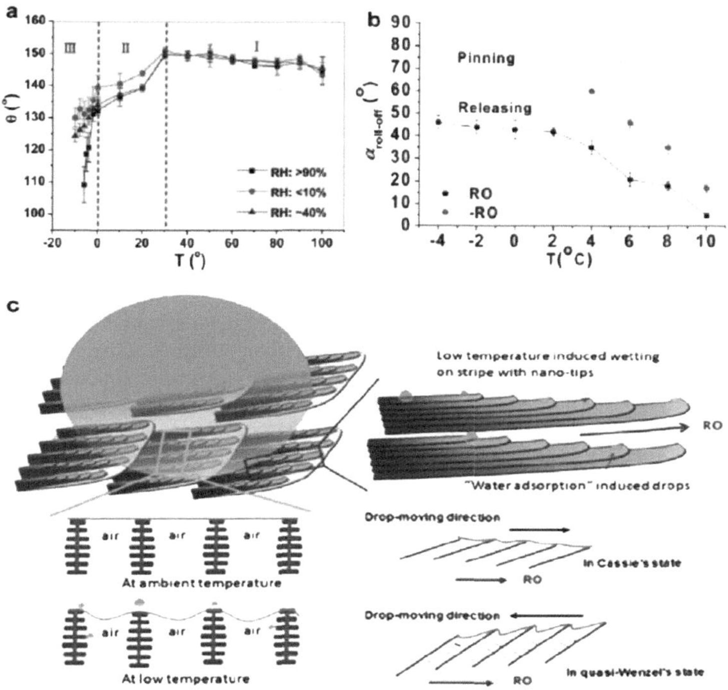

Figure 2.10 (a) The contact angles (θ) of water versus the temperatures (T) under different relative humidity (RH). (b) The anisotropy of rolling-off angles ($\alpha_{\text{roll-off}}$) of drops at temperatures (T). (c) Illustration of low temperature wetting and wetting anisotropy of butterfly wing.

Environmental scanning electron microscopy (ESEM) shows the superhydrophobicity of butterfly wings at microlevel. For example, the microdroplets are formed on nanostripe of scale on wing (temperature of 2°C, humidity of 100%). Figure 2.11, left shows the water condensed droplets on surface of overlapping scales on *M. rhotcnor* butterfly wing. The right shows the water condensed droplets on surface of overlapping scales on *M. aega* butterfly wing. The several micrometer sized droplets are suspended up on nanostructures of scales. It indicates that butterfly wings display strong superhydrophobicity at micro-/nanolevel. The observation further demonstrates that the surface of butterfly wings has low-temperature water repellency and also robust anti-fogging functions.

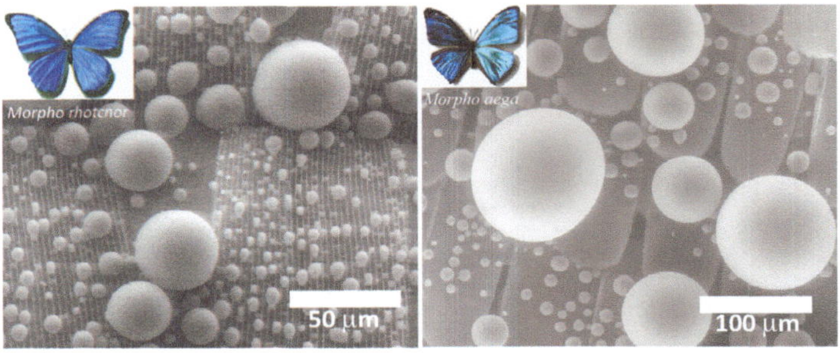

Figure 2.11 The left image shows the water condensed droplets on surface of overlapping scales on *M. rhotcnor* butterfly wing. The right image shows the water condensed droplets on surface of overlapping scales on *M. aega* butterfly wing.

2.2 Artificial Surface and Functions

Inspired by the fantastic properties of natural creatures, biomimetic fabrication of artificial materials with functional performance similar to or even better than that of nature has become a hot topic in both fundamental science and industrial applications in the past few years.[7,25,26] In order to reproduce the structures, novel processing methods have been adopted, as illustrated in Fig. 2.12, usually there are two typical kinds of surfaces: the 2D anisotropic structured surface (left) and 1D asymmetric structured surface (right).

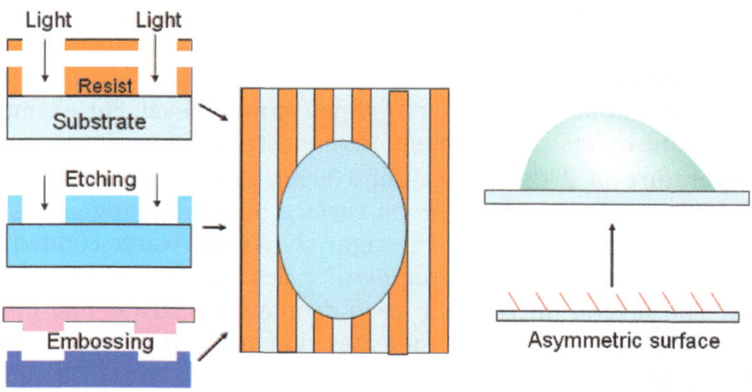

Figure 2.12 Typical fabrication approaches to form the anisotropic wetting surface.[27]

2.2.1 Optical Lithography and Interference Lithography

Optical lithography creates patterns on substrates via photochemical reactions with an appropriate light source, a mask, and a photoresist (PR). Optical lithography could produce patterned line surface structures with line widths ranging between several to hundreds of micrometers. Moreover, this technique could be used to ultimately produce either chemical patterns or topographical patterns for studying anisotropic wetting behavior. Bliznyuk et al.[28] studied the scaling of anisotropic droplet shape on chemically patterned surfaces, which were generated via optical lithography and contained micrometer-scale striped patterns of alternating hydrophilic SiO_2 and hydrophobic fluoroalkylsilane self-assembled monolayers. They found significant disparity of contact angles in directions parallel and perpendicular to the stripes. Provided the droplet dimensions were approximately 1–2 orders of magnitude larger than the width of the stripes, it is also reported that droplet shape and accompanying contact angles depended on the relative width of the hydrophobic and hydrophilic lines rather than the absolute line pattern size. They concluded that anisotropic droplets may exist both on chemically patterned surfaces and topographic patterned surfaces.

Besides chemically patterned surfaces, optical lithography may also aid in ultimately producing topographic patterned surfaces. In one example, Sommers et al.[29] generated parallel micrometer-scale grooves on aluminum surfaces via standard optical lithography, which practices were employed in preparing the plates for etching. Positive PRs, S1813 and AZ4620, are selected to mask the surface, and a mixture of BCl_3 and Cl_2 gases is used to perform the etching. They demonstrate that the peel strength of this adhesive is dependent on the amount of overhang of a thin flexible cap on the top of each fiber. By precisely placing the center of this cap offset to the center of the supporting post, the peel strength of the adhesive can be altered when pulled off in different directions. This study effectively illustrates that anisotropic surface morphology alone may affect droplet geometries without requiring chemical surface modifications.

Advancements in nanotechnology provide a useful route for studying anisotropic wetting on submicro-scale surfaces. The interference lithography (IL) approach has also been used to generate grooved biomimetic surfaces to investigate anisotropic wetting. Xia et al.[30] used IL to produce polymer patterns. A developable bottom anti-reflective coating layer was spun on the precleaned sample before applying PR. Both positive-tone PR and negative-tone PR were used. Parallel PR/wet lines (1D) were formed on the Si wafer surface after exposing, baking, and developing. Modified 1D surfaces were prepared with silica nanoparticle colloidal dispersions using spin coating at 4000 rpm for 30 s. Also, some samples with 1D PR patterns were prepared atop a thin silica nanoparticle film deposited with spin-coating. Both the degree of anisotropy and the droplet distortion are extremely high as compared with previous reports because of the well-defined nanostructural morphology. Wu et al.[31] employed an advanced IL approach to produce micropearl arrays for regulating 2D anisotropic wetting. The resin was spin-coated on a glass slide at different rotation speeds to obtain the needed thickness of the resin. The improved IL was based on intensity modulation of four beams. The four beams were overlapped on the sample to produce designed interference patterns. After irradiation, micro pearl arrays were formed. They demonstrated systematic regulation of contact angles (CAs) on anisotropic surfaces by adjusting the resin thickness and the laser beam intensity ratio (Fig. 2.13). Wu et al.[32] evaluated surfaces prepared via IL that contained submicrometer and low-micrometer grooves of different line widths, periods, and heights and found that the anisotropic wetting strongly depended on the height, and weakly on the about the magnitude of several micrometers.

To overcome the limitation of straight features, inclined lithography (IncL) was developed. IncL enables the formation of 3D geometries such as inclined micropillars and truncated cones by tilting the wafer with respect to the light source and optionally rotating the wafer and/or the mask during UV exposure. For example, Sato et al.[33] reported that after bonding the SU-8 top cover and the SU-8 bottom substrate, the micromesh and microchannel structures were defined by using the combination of the inclined multiangle backside exposure and the front-side exposure. As a result, 3D

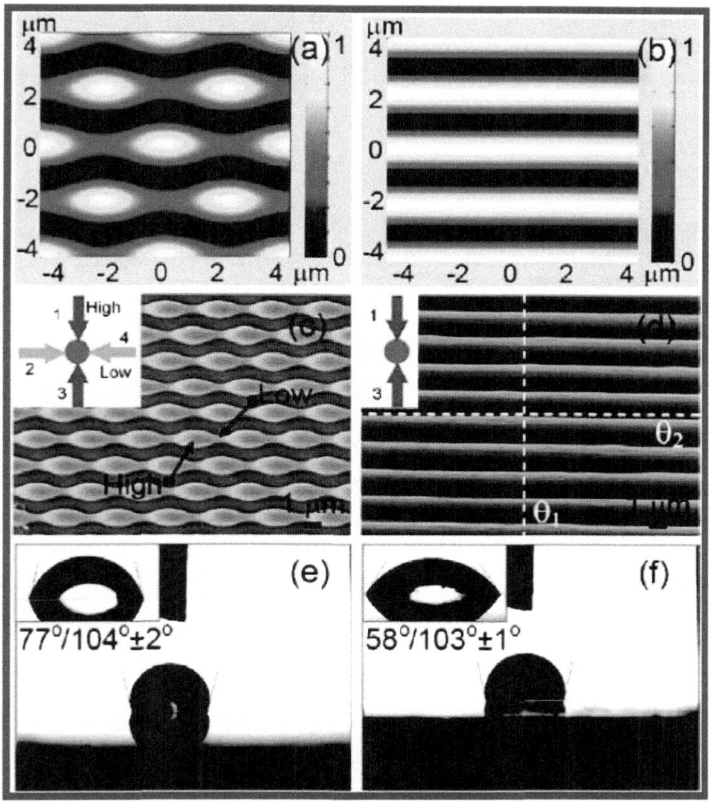

Figure 2.13 Micropearl arrays fabricated by improved laser IL. (a & b) Calculated intensity distributions of modified four-beam and common two-beam IL. (c & d) SEM images of micropearl arrays and grooved structures prepared by improved four-beam and common two-beam IL. (e & f) CA measurement.[31]

fine micromesh structures were seamlessly embedded in the SU-8 sealed microchannel (Fig. 2.14). The main limitation of anisotropic structures fabricated by lithography remains the use of the PR as a functional material. For example, the mechanical, electrical, and thermal properties of most PR are inferior to the properties of metals, silicon, or other inorganic materials. Nonetheless, this limitation can be overcome via coating or infiltration of polymer structures with inorganic films or nanoparticles.

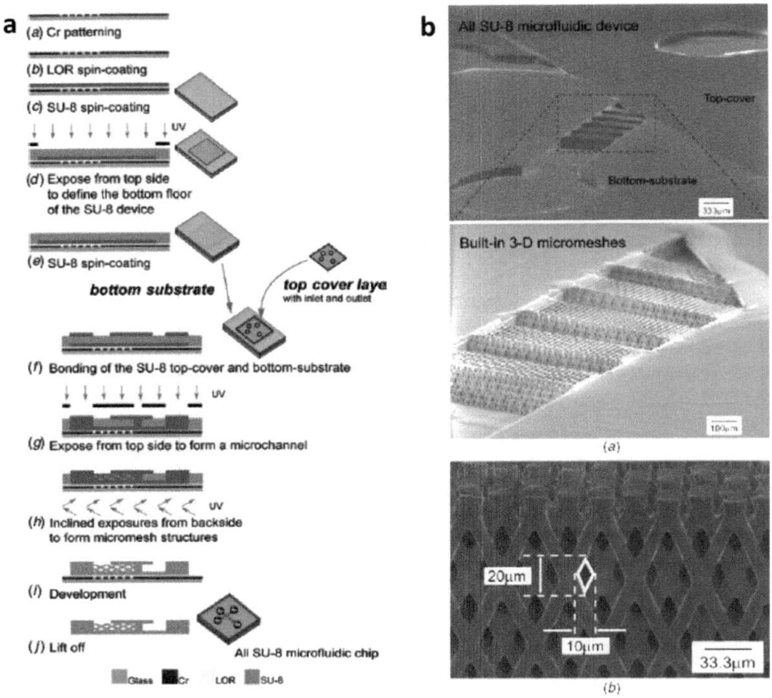

Figure 2.14 (a) Fabrication process of an all SU-8 microfluidic chip with a built-in 3D micromesh structure. (b) SEM photomicrograph of the all SU-8 microfluidic chip.[33]

2.2.2 Femtosecond Laser and Etching

Femtosecond lasers are a quick and efficient approach for producing anisotropic microstructures on a variety of materials without requiring expensive masks and clean-room facilities. Using a precise computer controlled stage could enable rapid and precise micromachining. In one study, Zhang et al.[34] reported that microscale hierarchical structures consisting of parallel grooves decorated by embossed triangle patterns are prepared by femtosecond laser irradiation on silicon wafers. The results show that increasing the vertex angle of the triangle and groove spacing could lead to the enhancement of wettability and anisotropy, respectively. In another example, Chen et al.[35] used a femtosecond laser to fabricate a Si surface with anisotropic wetting behavior due to adhesive forces of these structured dominated by Si.

Laser etching is another method used for patterning substrates. For example, using a single-step laser etching, Gao et al.[36] obtained micropatterns with rugged nanoprotrusions on Si surface. The results showed that anisotropy would vary with the volume-induced pinning–depinning–repinning behavior of the droplet. The adhesive force of the structure is proven to be the key factor giving rise to the anisotropy wetting. Plasma etching also serves as an effective method for generating patterns on Si surfaces. Plasma etching permits morphological control of nanoscale or microscale features on Si surfaces. For example, Wang et al.[37] fabricated silicon elliptical pillar arrays with unique anisotropic wetting characteristics via plasma etching method. Due to the anisotropic morphology of the elliptical pillar, the Si elliptical pillar arrays show anisotropic wetting property, which could be enhanced by increasing the elliptical pillar aspect ratio (AR). To mimic rice leaf surfaces, Zhang et al.[38] also used a similar approach to fabricate anisotropic, hydrophobic surfaces containing patterned silicon cone arrays. Wu et al.[39] reported an interesting process to form oblique nanowire (NW) array by etching. At first, an insulating polymer film was covered with a layer of discontinuous Au granular. Under Ar^+ ion bombardment, these Au nanoparticles act as a shadow mask to protect the polymer film underneath, which leads to formation of hilly structures on the polymer film. After a period of Ar^+ ion bombardment, the Au nanoparticles disappear and the hilly structure is positively charged. When successive Ar^+ ions move through the hilly structure, they will be repulsed by hilly structure. Because of the deflecting effect, the top of the hilly structure will be bombarded off more slowly than its bottom part, and NWs are formed (Fig. 2.15).

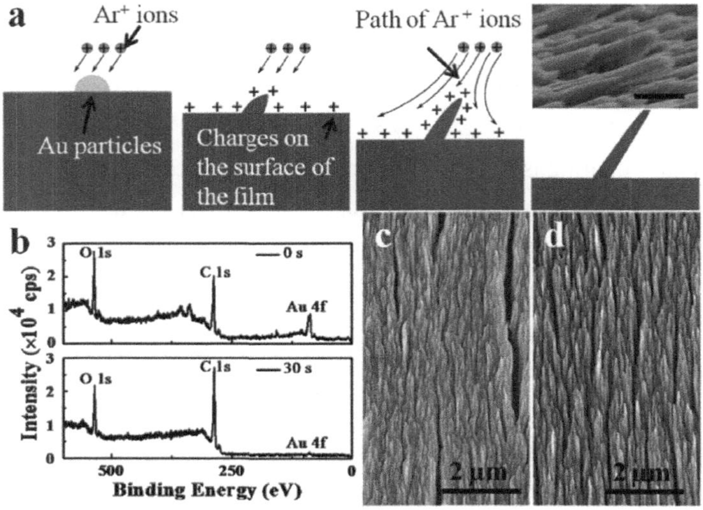

Figure 2.15 Formation mechanism of NWs. (a) Schematic of oblique NW formation. Inset: A typical SEM image of oblique poly(ethylene terephthalate) (PET) NWs; scale bar represents 2 μm. (b) X-ray photoelectron spectroscopy (XPS) spectra of PET sheet coated with Au film before (upper panel) and after 30 s etching (lower panel). (c) poly(3,4-ethylenedioxythiophene) (PEDOT) oblique NWs on Si substrate. (d) PEDOT oblique NWs on PET thin film. The average thickness of Au is 4 Å and the bombarding time is 1 h.

2.2.3 Embossing, Imprinting and Replica Molding

As compared to lithography, embossing, imprinting, and replica molding methods generate patterned surfaces in a facile manner without requiring clean-room operations. Typically, a soft polymer such as polydimethylsiloxane (PDMS) or polyurethane could be used to cast a negative of a hard master mold. Then, the negative is used to cast a replica. In contrast to nanoimprint lithography (NIL), embossing, imprinting, and replica molding can produce surface features over a wide range of length scales and allows a wide variety of replica materials and geometries to be made. Additionally, the match between the master and the subsequent transfer materials is determined largely by the physical and chemical compatibility of the materials. Neuhaus et al.[40] prepared poly(ethylenealt-tetrafluoroethylene) samples with hot-embossed periodic grooves or squares as follows: stamps with the desired structures were produced using standard photolithography and reactive-ion etching. Ethylene-tetra-fluoro-ethylene (ETFE) foils of 100 μm thick were placed between the stamp and a polished silicon wafer. After heating to 220°C, a pressure of 4 MPa was applied for 20 min. The system was cooled to room temperature before the pressure was released. The assembly was then removed from the press and separated using a razor blade. Liu et al.[41] reported that a surface embossing technique was applied to fabricate microgrooved aluminum fin stocks in a mass production manner. Embossing plates with the desired roughness design by laser treatment were first made. Pitches of the grooves and engraving depth can be adjusted. The embossing plates were simply pressed against the fin stock, and the generated profile was successfully transferred onto the aluminum surface. Yang et al.[42] used a hot-embossing method to produce grooved poly(methyl methacrylate) (PMMA) surfaces for studying the effect of sessile droplet volume on wetting anisotropy. First, photolithography and reactive ion etching (RIE) were employed to generate silicon masters containing specific topography. Using suitable pressure and temperature, the patterns of the silicon masters were transferred to the PMMA sheets and subsequently the surfaces were treated via plasma polymerization. Their results showed that for both hydrophobic and hydrophilic surfaces, anisotropic wetting on these grooved surfaces increased with larger drop size.

Aiming to a straightforward use of the natural templates, bio-inspired anisotropic micro-/nanosurfaces have been fabricated successfully by utilizing the outward wings of grasshoppers (OWGs) as a natural stamp by Zhang et al.[43] The as-prepared PDMS surfaces not only replicated the patterned microstructure of OWG well, but also exhibited anisotropic high hydrophobicity (Fig. 2.16). Its CA with a 2 μL water droplet was 137.8° measured from the parallel direction to the ridges and 109.5° measured from the perpendicular direction. In another study, elastomeric replicas of green have been produced by replica molding with PDMS.[44] The PDMS prepolymer was poured onto the surface of the green and thermally polymerized in situ at room temperature for 24 h. Polymeric stamps from the green leaf showed an arrangement of periodic microridges/microgrooves that induced anisotropic wetting, with a water contact angle variation of about 21% along the two principal directions. Using rice leaves as templates, Au surfaces with positive and negative biomimetic rice leaf textures were also fabricated.[45] Micro-/nanoscale hierarchical structures on the surface were anisotropic and showed the anisotropic sliding angle performances. Huang et al.[46] used the natural butterfly wings as the templates to replicate the micro- and nanometer scale hierarchical photonic structures via an alumina coating through a low-temperature atomic layer deposition process. The results indicated that the alumina replicas possessed the inherited photonic properties of the natural wing scales and other novel properties.

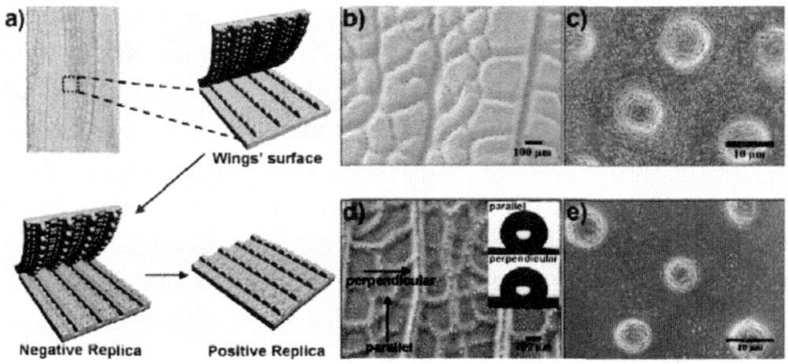

Figure 2.16 (a) Schematic representation of the two-step replication process. (b & c) SEM images of first-step PDMS replica of OWG. (d & e) SEM images of second-step PDMS replica of OWG. Inserts in (d) Anisotropic high hydrophobicity.[43]

To obtain artificially oriented ratchet surface, the PDMS prepolymer was used as the first-step replica of the negative microstructure (e.g., the taper-ratchet array) on ryegrass leaf, and polyvinylidene fluoride (PVDF) was used as the second-step replica of the positive microstructure. Due to the oriented structures, the droplet is easily driven to move directionally via exerting the vibration onto the surface (Fig. 2.17).

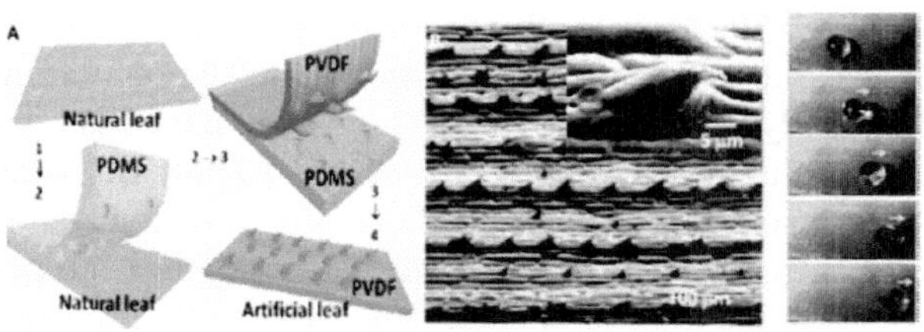

Figure 2.17 (a) Fabrication of artificial structured surface. (b) SEM image of the artificial surface. The inset is a single taper-ratchet with the tip oriented direction. (c) When the surface is exerted by 50 Hz vibration, the drop of 10 μL rolls off along the oriented direction.

The artificial surfaces are composed of taper-ratchets with tips, the tips have the tilted angle (ε), apex angle (δ), and gradient width on upper of ratchets (as illustrated in Fig. 2.18), forming asymmetric gradient array. The asymmetric oriented taper-ratchet dependence on direction modulates the style of three-phase contact angle, accordingly, generates the asymmetric retentions and modulates the behavior of drop in the 1 and 2 direction. So the tips of taper-ratchet are thought to be mainly oriented to modulate the behavior of wetting in directions, and also to change the advancing and receding angles in alternative 1 and 2 directions. In the 2 direction, the oriented tips of taper-ratchet oppose the moving of liquid due to the tilted-up angle, and change greatly the type of TCL, accompanying with the advancing angle $\theta a_2 = \theta a_0 - \varepsilon$ and receding angle $\theta r_2 = \pi - \theta r_0 - (\varepsilon + \delta)$, respectively. But in the 1 direction, the oriented tips of taper-ratchets obey the moving of liquid and the TCL is little changed, accompanying with advancing angle $\theta a_1 = \theta a_0 - \varepsilon$ and receding angle $\theta r_1 = \theta r_0 + \varepsilon$, respectively. In addition, the tips of taper-ratchets form the array in gradient diameter. The solid–liquid contact width (w) forms gradient along the tips of taper-ratchets, generating the release of liquid in the 1 direction and pinning of liquid in the 2 direction. For the single tip of taper-ratchet, there are maximum contact width ($w_{0,\max}$) in the 2 direction and minimum contact width ($w_{0,\min}$) in the 1 direction. These features induce the special retention force ($f_{1,2}$) described as following:

$$f_{1,2} \sim 2w_{1,2}\gamma\left[\sin\left(\frac{\theta_{a_0}+\theta_{r_0}}{2}-\beta_{1,2}\right)\sin\left(\frac{\Delta\theta_0}{2}-\chi_{1,2}\right)\right] \tag{1}$$

Where θa_0 is intrinsic advancing angle; θr_0 is intrinsic receding angle, and $\Delta\theta_0$ is difference of intrinsic advancing angle and intrinsic receding angle, that is, $\Delta\theta_0 = \theta a_0 - \theta r_0$ (herein, $\theta a_0 = \sim 100°$ and $\theta r_0 = \sim 55°$ for smooth PVDF surface). Among which $\beta_1 = 0$; $\beta_2 = (\pi - \delta)/2$ and $\chi_1 = \varepsilon$; $\chi_2 = \varepsilon - (\pi - \delta)/2$; the subscripts "1, 2" indicate the cases along the 1, 2 direction, respectively. The $w_{1,2}$ indicates the solid–liquid contact width in 1, 2 direction, respectively, and for direction moving of drop, $w_2 > w_1$. The w_2/w_1 can be estimated to be 25 at maximum and the f_2/f_1 from the Eq. (1) is up to 100 at maximum force ratio to be contributed to the unidirectional water shedding-off.

Such a gradient retention force in alternative direction helps to suspend the liquid meniscus and bridge on the top of ratchet by the decreasing contact width along the taper tip in the 1 direction. It makes drop-easy rolling-off, generating the release of liquid (Fig. 2.18b). Otherwise, it allows the liquid meniscus to wet the tip of taper-ratchet in depth during the increase of contact width along the 2 direction. It makes drop to be in metastable state, and is reversibly pinned by the tips of taper-ratchets, and generates the pinning of liquid. As a result, the taper-ratchet surface enables drop easily to be driven along the oriented direction of tip for the unidirectional water shedding-off (i.e., in the 1 direction), rather than in the 2 direction.

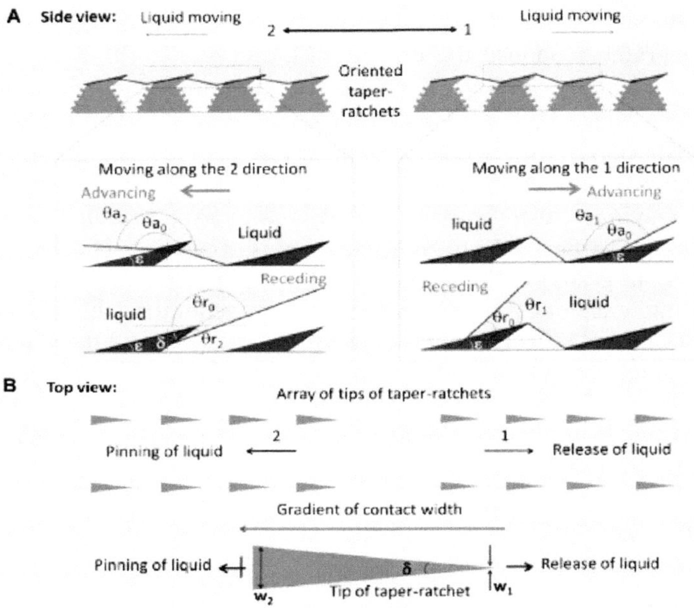

Figure 2.18 Illustration on directional shedding-off. (A) Side view. The tip of taper-ratchet has tilted-up angle (ε) and apex angle (δ). The 1 direction is along the oriented tip but the 2 direction is against the oriented tip. The left inset: In liquid moving along the 2 direction, there are advancing angle $\theta a_2 = \theta a_0 - \varepsilon$ and receding angle $\theta r_2 = \pi - \theta r_0 - (\varepsilon + \delta)$. The right inset: In the liquid moving along the 1 direction, there are advancing angle $\theta a_1 = \theta a_0 - \varepsilon$ and receding angle $\theta r_1 = \theta r_0 + \varepsilon$. (B) Top view. The tip generates a gradient retention force resulted from the gradient of contact width ($w_2 \rightarrow w_1$) at solid–liquid interfaces along from 2 direction to 1 direction.

Inspired from the wing-locking device of a beetle, the silicon masters with micro- and nanoholes were prepared by photolithography and subsequent RIE by Pang et al.[47] The masters were firstly treated with a fluorinated self-assembled monolayer and then annealed at 120°C for 20 min. After drops of UV-curable polyurethane (PU) elastomer or soft polyurethane acrylate (PUA) prepolymers were dispensed onto the master, a flexible polyethylene terephthalate film was slightly pressed against the liquid drop to be used as a supporting backplane. After preparing a polymer replica by UV exposure and mold removal, the PUA replica was additionally exposed to UV for several hours to cure. To mimic wing-locking device of a beetle, they fabricated various high AR, vertical micro- and nanopillars on a flexible substrate and investigated the shear locking force with different preloads.

As an imprinting technique, NIL is good choice to fabricate patterned surfaces. NIL was invented by Chou and colleagues in 1995. NIL uses a rigid master mold that is pressed into a soft polymer. Nanoimprinted polymer surfaces can also be used as masks for patterning the underlying substrate, including metal, semiconductor, and oxide thin films. Therefore, NIL has been used in many fabrication processes, such as organic electronics, biosensors, and nanofluidics.[48] In order to fabricate anisotropic biological structures, Zhang et al.[49] used sequential imprinting based on conventional NIL to generate hierarchical structures on polystyrene (PS) and PMMA films. All the treated molds first underwent a self-cleaning imprint on PS or PMMA thin films to further remove the physisorbed silane that escaped sonication; the self-cleaning imprints were made at 120°C and 40 bar for 300 s. The cleaned molds were then used to carry out imprinting on samples for wetting property studies. PMMA showed an anisotropic wettability from 6° to 54° and PS had an anisotropic wettability from 8° to 54°.

2.2.4 Wrinkling and Directed Mechanical Deformation

Wrinkles result from stress gradients or property mismatch across the thickness of planar materials and are ubiquitous in nature. Many studies have shown that wrinkles can be formed with scales ranging from nanometers to millimeters due to mechanical instabilities. One typical example involves the formation of microwrinkles on PDMS via mechanical stretching and releasing of oxide/PDMS bilayers. Contact between a soft film and a flat rigid surface generate instabilities that result in corrugation of the PDMS surface. In one study, a sheet of cross-linked PDMS was mounted on a strain stage and subsequently stretched uniaxially to a strain (Fig. 2.19).[50] Ultraviolet-ozone was used to oxidize the PDMS surface forming a thin silicate layer. Owing to modulus mismatch, formation of 1D wrinkles occurred with uniaxial release of the applied strains. The wavelength and the amplitude of the wrinkles could be controlled by manipulating the properties both of the stiff film and the substrate. A liquid droplet upon such microwrinkled surfaces often exhibits anisotropic wetting.

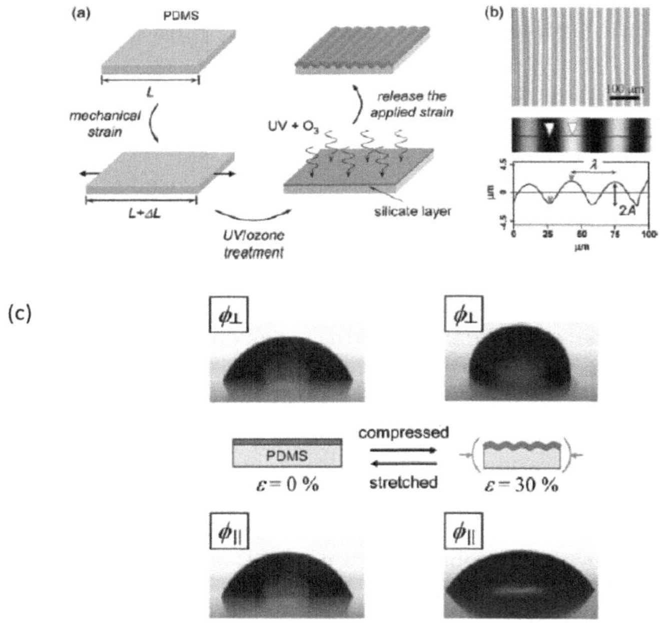

Figure 2.19 (a) Schematic illustration of steps for generating sinusoidally microwrinkled patterns on PDMS substrates. (b) Optical micrograph of wavy patterns on a PDMS substrate and top view atomic force microscopy (AFM) image of the same sample and its corresponding surface height profile plotted along the line in black. (c) Images of water droplet shape in two views perpendicular and parallel to the direction of the grooves.

Based on well-established theories of structure buckling and solid–liquid contact, Zhang et al.[51] showed that this objective may be achieved via a hierarchically wrinkled surface. An *n*-level wavy surface produced via sequential mechanical buckling upon release of prestrains in a multilayer system, was used as a model system to investigate the relationship among mechanical strain, surface morphology, wettability, and adhesion on hierarchical surfaces. The results indicate that the surface can be mechanically switched between superhydrophilicity and superhydrophobicity. It is shown that such a surface possesses multiple function states, including SH, hydrophobic, hydrophilic, and superhydrophilic states associated with wettability, as well as weak and strong sticky states associated with adhesion. All these functional states can be repeatedly and reversibly switched in a pure mechanical manner, by applying different levels of tensile strain on substrate.

Arrays of bent or inclined pillars can also be fabricated by applying asymmetric stresses to cause straight pillars bending. In the first fabrication step, vertical micro- or nanopillars are fabricated by etching or molding from a master template. Then, the pillars are subsequently bent by directional metal deposition, electrochemical coating, e-beam irradiation, thermal treatment, or a combination of these methods. This step induces a stress gradient across each pillar causing bending due to a mechanical effect analogous to the thermal actuation of a bimetallic strip. These anisotropic arrays have found applications in unidirectional wetting, anisotropic adhesion, chemical and biological sensing, and actuation.[52,53] Kim et al.[52] reported on unidirectional wetting and spreading of a water droplet on stooped polymer nanohairs fabricated by replica molding and oblique electron beam irradiation. First, high AR vertical polymer nanohairs were generated by replica molding. Subsequently, the as-prepared vertical nanohairs were coated with 4 nm thick platinum to prevent electron charging and to achieve the appropriate surface wettability. The samples were obliquely exposed to e-beam and irradiated for 20 s under vacuum using SEM, resulting in stooped nanohairs from asymmetric decomposition, and shrinkage of the polymer. When a droplet was placed on the surface with directionally bent nanohairs, it wet and spontaneously spread over the surface in a unidirectional manner along the stooped direction with one edge being pinned to the substrate.

2.2.5 Other Methods

There are many other methods for fabricating the anisotropic wetting structures, such as self-assembly, metal assisted chemical etching, glancing angle deposition, oblique angle polymerization, and oriented nanostructure growth.[27] Balu et al.[54] patterned SH paper substrates with high surface-energy black ink using commercially available desktop printing technology. The patterns were designed using standard word processing software. Experimental data on the adhesion forces showed good agreement with classical models for "drag" and "extensional" adhesion. Prakash et al.[55] proceed by presenting a new approach to fabrication of artificial unidirectional SH surfaces exploiting the mechanics of peeling (Fig. 2.20). The technique relied on capillary bridges to form and break between two polymer films, thus allowing microscopic pillars with controlled directionality to be formed on very large surfaces. The process is also suitable for implementation in a roll-to-roll process for producing large sheets in a manufacturing setting. In this method, polystyerene beads were attached to a glass plate with an epoxy adhesive in a random pattern. The thermoplastic sheet was attached to a flexible substrate and mounted on a metal roller. The system was heated to 125°C and the roller was translated by a motorized stage at a set velocity of 5 mm/s. The net diameter of the roller governs the tilt angle of the artificial hairs formed on the surface. The contact of each bead with the sheet resulted in a capillary bridge. Doubtless, with the development of technology, more and more methods and routes would emerge to fabricate more accurate anisotropic structures.

2.3 Potential Applications

The ability to control specific wettability of a solid surface is of universal importance in academic sciences and industrial technologies. Among these, the research on the topic of bio-inspired surfaces with anisotropic wettability is a large and growing field.[25,26,32,47,56] Biomimetic research indicates that the cooperation of the unique structural design of these biological materials or organisms with intrinsic material properties plays a crucial role to achieve the anisotropic wettability. Although investigations

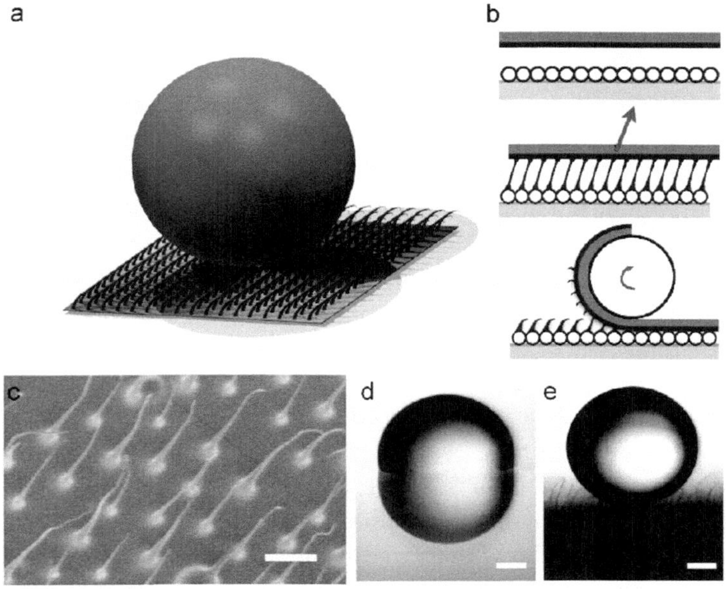

Figure 2.20 (a) An idealized unidirectional SH surface based on the form of water-walking arthropod cuticle. (b) Schematic illustration of the pattern peeling technique used to fabricate such surfaces. (c) SEM photomicrograph of a section of tilted pillars fabricated by peeling a polyester-based thermosetting adhesive film off a bed of PS. Water drops sitting on (d) a flat sheet of the silanized polymer film and (e) the polymer film textured by pattern peeling. Scalebars: (c) 200 μm; (d) 500 μm; (e) 500 μm.[55]

involving anisotropic wetting are in the early stages of research and the practical application of nature-inspired surfaces is still limited, a lot of effort is gradually being accumulated, providing a great opportunity for the development of various laboratorial and industrial products, involving biomedical devices, self-cleaning, and fluidic control and transport in microfluidic devices.[13,26,37,51] Inspired by natural directional surfaces, engineers have developed a myriad of synthetic surfaces with precisely tuned physicochemical properties to transport water droplets and soft materials, control liquid spreading, provide directional adhesion, and exhibit directional friction.

2.3.1 Anisotropic Wetting and Unidirectional Spreading

Anisotropic wetting is phenomena in which apparent contact angles on a solid surface are related to the direction of view. Recent studies have demonstrated elongated droplets, unidirectional spreading of liquids. These anisotropic surface architectures have several potential applications including open-channel microfluidics and self-cleaning surfaces. Drelich et al.[57] fabricated an anisotropic wetting surface by contact printing of 2.5 μm wide stripes of self-assembled monolayers of hexadecanethiol and mercaptohexadecanoic acid. Consistently, the contact angles from the view parallel to the stripes were 2–10° larger than the view perpendicular to the stripes. In other examples of chemically patterned anisotropic surfaces, uniform 200 nm wide hydrophilic mica channels were separated by 800 nm wide hydrophobic stripes of L-α-dipalmitoyl-phosphatidycholine monolyers.[58] The monolayer stripes were formed by an oscillating meniscus during the withdrawal of the sample from the organic monolayer solution. The anisotropic properties of these channels were demonstrated as templates for selective adsorption from solution. For example, when these substrates were withdrawn from colloidal solutions, only the hydrophilic channels were coated with gold particles, dye, and $FeCl_3$ by simple capillary filling. Further, the anisotropic wetting surfaces made by chemical patterning can be used to control the alignment and placement of 1D nanostructures, such as single-walled carbon nanotubes (SWCNTs). Sharma et al.[59] used micro contact printing to pattern thin stripes of self-assembled monolayers of alternating hydrophilic/hydrophobic surface chemistries. Upon, cylindrical droplets were formed on the hydrophilic stripes after SWCNT aqueous solutions wetted the chemically patterned surfaces. The flow inside these droplets during evaporation led to the controlled alignment and placement of individual SWCNTs along the middle of the hydrophilic stripes. Liu et al.[41] used a surface embossing technique to impart micro-grooved topographical features on an aluminum fin stock to enhance water drainage. It was found that the microgroove structure caused anisotropic wetting behavior of the droplets, and apparent contact angles along the microgrooves ($\vartheta_{\perp}$) were always larger than those measured from the other direction ($\vartheta_{||}$). A consistent reduction of the critical sliding angle was observed and explained in terms of contact line discontinuities and contact line pinning effects. Due to the roughness anisotropy, water droplets exhibited an elongated shape along the direction of microgrooves. These results showed the promise of water drainage enhancement by changing surface morphology and smaller groove pacing, larger depth, and steeper groove sidewalls were favorable for drainage enhancement.

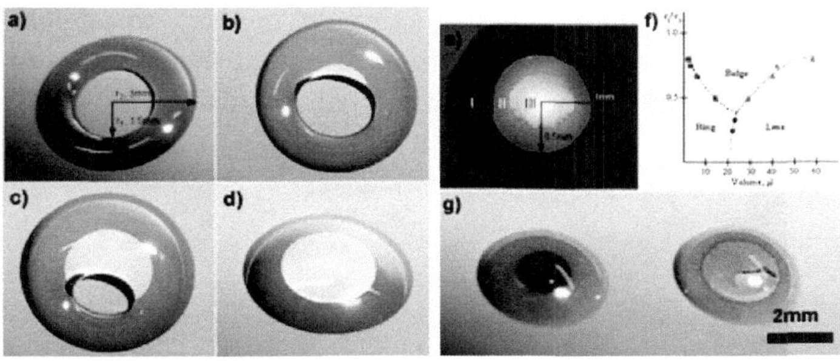

Figure 2.21 Droplet morphologies on a ring-shaped completely wetting domain surrounded by ultrahydrophobic surface.[60]

Beyond merely producing a simplified elongated droplet due to anisotropic wetting on bulk patterned surfaces, it is possible to create more complex shaped droplets using a combination of prepatterned areas fabricated by some processes and anisotropic wetting areas. For instance, complex droplets such as rings, bulges, and lenses were formed on chemically modified Si nanograss using patterned isotropic SH and superhydrophilic areas.[60] It is practical to form more complex patterns with droplet-evaporation–induced techniques and functional solution such as colloidal particle and carbon nanotube suspensions.[30] Such complex droplets hold potential for droplet microfluidics, such as droplet splitting, especially for bioapplications.

Unidirectional liquid spreading is a special case of anisotropic wetting where liquid droplets are pinned in three directions and wet the surface in only one direction.[52,61] For the view perpendicular to the spreading, the contact angle is symmetric; and for the view parallel to the spreading direction, there are two distinct contact angles, pinned and spreading. Chu et al.[61] designed asymmetric nanostructured surfaces to achieve unidirectional liquid spreading, where the liquid propagates in a single preferred direction and pins in all others. They used photolithography and deep RIE to fabricate an asymmetric nanostructured surface comprising Si pillar arrays. These Si nanopillars deflected due to high residual stresses from gold that was deposited on a single side of the pillar. Evidence of the unidirectional response of a water droplet atop nanoscale features is nicely observed. The droplet will propagate in a single direction on bent nanopillar structured surfaces. This study indicated that factors including the degree of nanostructure asymmetry, the height-to-spacing ratio of the nanostructures and the intrinsic contact angle contributed to the spreading characteristics of droplets.

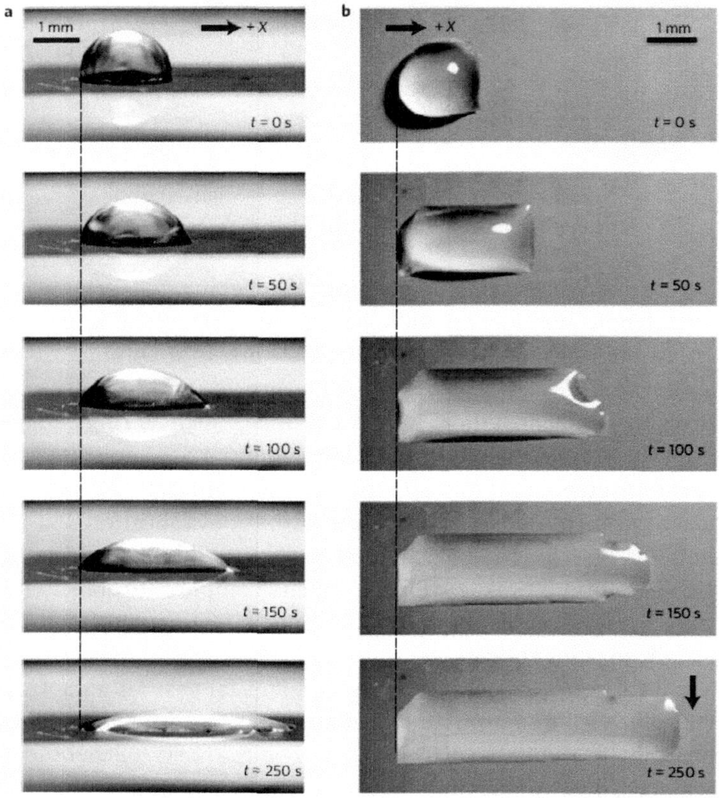

Figure 2.22 Scanning electron micrographs with uniform arrays of asymmetric nanostructured surfaces. Time-lapse images of uni-directional spreading of a liquid droplet.[61]

Mele et al.[44] demonstrated that the leaves of *Strelitzia reginae* were used to produce artificial surfaces with different wetting properties. The imposed multiscale microstructures endowed the bio-inspired surfaces with anisotropic wettability and superhydrophobicity. The morphology favored water spreading in the direction parallel to the microfeatures and liquid trapping in the perpendicular direction, determining anisotropic wetting with a water CA variation of about 21% along the two directions. This anisotropic wettability has important potential applications in microfluidic control and directional water transport. In one example, Kim et al.[52] demonstrated unidirectional wetting upon surfaces containing stooped polymer nanohairs with ARs of ~10. The nanohairs were fabricated via replica molding and oblique electron beam irradiation in a geometry-controllable manner. It was observed that droplets spread spontaneously and unidirectionally along the stooped direction of the nanohairs over time, not in perpendicular direction. The reason that the droplet spread predominately along the stooped direction was that the forward direction had a much lower energy barrier than the reverse direction. Finally, the droplet reached equilibrium and stable state. In order to study the influence of surface chemistry on the wetting of structured surfaces, Neuhaus et al.[40] produced microstructures consisting of grooves or squares via hot embossing of poly(ethylene-alt-tetrafluoroethylene) ETFE substrates. The structured substrates were modified with polymer brushes, thereby changing their surface functionality and wettability. The results indicated that water droplets were most strongly pinned to the structure when the surface was moderately hydrophilic. As a result, the droplet shape was determined by the features of the microstructure. The water CA were considerably higher than on flat surfaces and differed, in the most extreme case, by 37° when measured on grooved substrates, parallel and perpendicular to the grooves. On hydrophobic substrates, the same effects were observed but were much less pronounced. On very hydrophilic samples, the microstructure had no influence on the drop shape. These findings were explained by significant differences in apparent and real CA at the relatively smooth edges of the embossed structures. Finally, the highly anisotropic grooved microstructure was combined with a gradient in polymer brush composition and wettability to achieve the directed spreading of water droplets could be observed.

In summary, the typical potential applications of anisotropic wetting include flow valves, microfluidics, biosensors, evaporation-driven pattern formation, droplet transport, microreactors, and tunable/switchable surfaces. Notably, we have not attempted to cover all potential applications of anisotropic wetting. For example, another important direction involves interaction of biological components with the anisotropic wetting surfaces for potential applications in biomedical devices. Few demonstrations for potential application in sensors, and microreactors have been performed so far. We expect future advancements of research findings and applications incorporating anisotropic wetting.

2.3.2 Adhesion

Anisotropic wetting adhesion is the phenomenon by which the roll-off angle of a liquid droplet is different along different directions. It is known that a droplet tends to slide on the surface of groove microstructures more easily along the direction parallel to the grooves rather than perpendicular. The use of anisotropic micro-/nanostructures that have anisotropic adhesive properties enables the locomotion of insects and reptiles on a variety of surfaces including vertical walls and even ceilings.[62,63] When an animal climbs on a vertical surface, the body weight generates a large downward force that needs to be supported by a strong adhesion of its feet. At the same time, the animal should be able to easily detach one or more feet for further locomotion without breaking the contact of the other feet.[64] This is achieved via anisotropic adhesives that can be released easily in one direction, while providing strong adhesion in another direction. In both insects and reptiles, such as gecko lizards, these adhesion pads typically consist of arrays of angled branching hierarchical fibers with spatula-shaped tips.[63] Because of the small contact size and large AR of the fibers, they are able to conform to both rough and smooth surfaces. In this way, millions of adhesive contact points are created between the foot and the climbing surface. The fibers adhere via van der Waals forces, possibly in combination with capillary forces.[62,65] The latter is a topic of debate. Lee et al.[66] have developed a facile approach for fabricating hierarchical nanohairs with high AR using a multibranched anodic aluminum oxide (AAO) template prepared by two-step anodization and barrier layer thinning processes. The formation of these structures was controlled using an AAO template with multilevel porous structures

(Fig. 2.23). Combined with nanohair yielding of a polymeric material during peeling-off from the hydrophobically modified AAO template, elongated hierarchical nanohairs with a high AR were fabricated without fiber collapse. These hierarchical nanohairs exhibited higher adhesion and frictional forces relative to the single level nanohairs and hierarchical nanohairs with a low AR, which are attributed to the maximized compliance of elongated hairs without lateral collapse. For geckos, some researchers reject the influence of capillary effect, while others report the humidity has an important influence on spatula adhesion.[65] Analysis of these natural structures indicates that adhesion anisotropy is often established by angled fibers, which yield a different tribological behavior along and against the pillar orientation.[67,68] This effect has been studied in detail in the feet of the green bush cricket[68] and the gecko,[62,69] which have fiber arrangements with an angle 30–70° to the surface. A different mechanism of adhesive anisotropy is observed in spider legs, where the contact area is composed of long setae that are only covered on one side with fine tips. Depending on the sliding direction of the counter surface, the spatula shaped tips are pushed onto the surface or bend away from it, hereby altering the adhesion properties.[70]

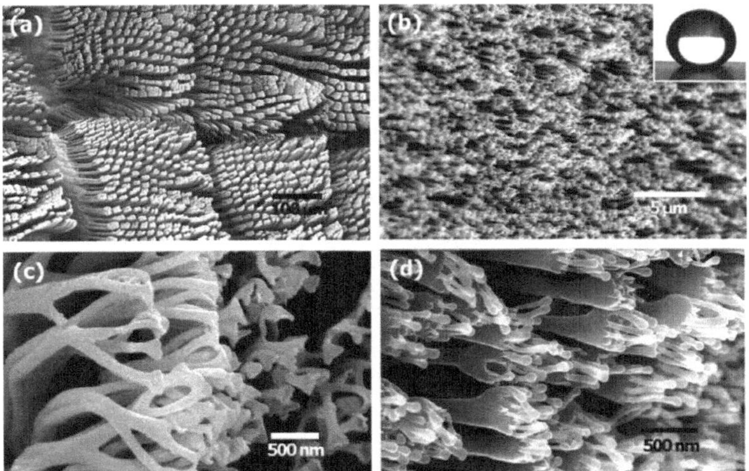

Figure 2.23 SEM images of a Tokay Gecko (*Gekko gecko*) (a and c) and fabricated hierarchical PS nanohairs with high AR (b and d).

Inspired by animals above, synthetic anisotropic dry adhesives are a recent engineering and ingenious fabrication methods have been developed.[69] Most engineered anisotropic dry adhesives utilize arrays of angled polymer microposts or nanofibers. These are created by flipping over vertical fibers by pressing them between heated rollers, by drawing,[71] or even by rupturing them in an angled direction. Arrays of angled fibers have been made in many methods, such as e-beam irradiation of an FE-SEM,[72] IncL,[64] and angled etching of a Si master mold.[73] The inclined structures can optionally be replicated at low cost by molding methods.[67,74] Anisotropy has also been introduced by coating one side of the pillars with a metal layer.[75] Depending on the shearing direction, the metal or the polymer is exposed to the target surface thereby introducing anisotropy. Anisotropic adhesive features have also been achieved via smaller nanostructures, such as by filling of ordered porous templates such as anodic alumina, and by chemical vapor deposition growth of carbon nanotubes (CNTs) or nanowires.[76] The performance of anisotropic adhesives is typically quantified by comparing adhesion in the forward and reverse lateral directions, and in some cases adhesion perpendicular to the target surface. For example, Zhang et al.[77] reported that butterfly wings ratchet structure was anticipated to generate an anisotropic adhesion between the superparamagnetic microdroplets and the surface. As a result, the droplets were deformed distinctly on the ratchet structured SH surface. Along one direction, the movement of the droplet moving appeared to be more difficult than along the other one. In contrast, no direction-dependent movement was observed on the SH surface without ratchet structure (Fig. 2.24).

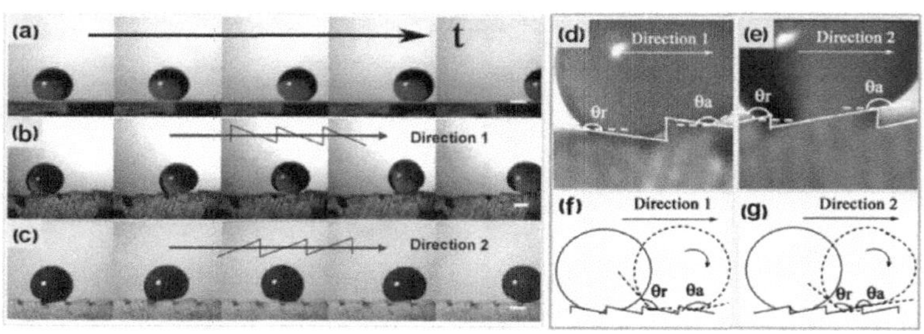

Figure 2.24 The time-sequence high-speed charge coupled device (CCD) images of microdroplets moving on the flat surface (a) and ratchet structured SH surfaces along directions 1 (b) and 2 (c), respectively. Scale bar: 1 mm. Magnified images of *M*-droplet that initially starts to move along magnetic field direction 1 (d) and 2 (e) vs. illustrations (f) and (g), respectively.[77]

Pang et al.[78] have presented a robust, reversible interlocking system using regularly arrayed, high-AR polymer fibers inspired from the wing-locking device of beetles (Fig. 2.25). Experimental and theoretical studies demonstrated that the interlocking was mediated by attractive van der Waals forces among high-AR hairy structures, resulting in a very high interlocking force in the shear direction and easy life-off in the normal direction. In contrast to other reversible binding systems, the current interlocking mechanism does not involve any complicated physical structures (e.g., hooks or loops) or surface chemical moieties, allowing for a simple, yet efficient, route to reversible interlocker in a noiseless and cost-effective manner. A simple theory was developed on the basis of force balance and hair-merging probability, which is capable of explaining the maximum shear adhesion and operating zones for various geometrical and material parameters. They also presented bidirectional, asymmetric interlocking behaviors between tilted micro- and nanohair arrays.[79] The measured shear adhesion force between two identical tilted microhair arrays turned out to be higher in the reverse direction than that in the angled direction, suggesting that the directionality of beetle's microtrichia may play a critical role in preventing the elytra from shifting along the middle of insect body. Furthermore, they observed dramatic enhancement of shear adhesion using asymmetric interlocking of various nanohair arrays.

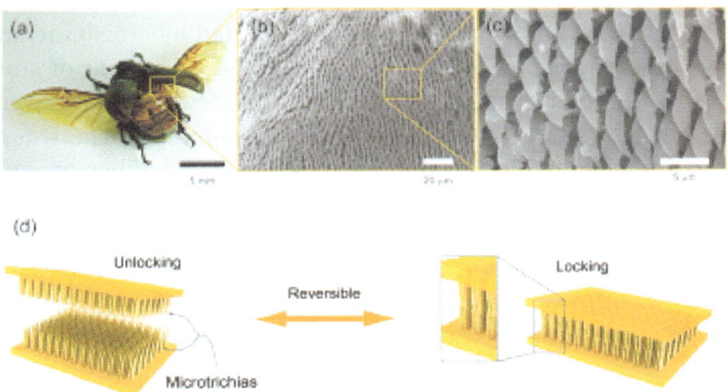

Figure 2.25 (a) Photographs of the wing-locking device of the beetle (*Promethis valgipes*). (b & c) SEM images of microtrichia on the cuticular surface with two different magnifications. (d) Schematic of folding and unfolding states of wing-locking device.[78]

Guo et al.[7] reported a novel taper-ratchet array on ryegrass leaf. It integrated a gradient of retention at solid–liquid interfaces in contrasting directions to reversibly generate the release or the pinning of solid–liquid contact lines, and accordingly, achieved effective directional water shedding-off properties. By mimicking taper-ratchets from ryegrass leaf, the polymer surfaces are fabricated successfully. They observed a robust property of directional water shedding-off. When external vibrations were added on polymer surfaces, the drops achieve a unidirectional self-shedding along the oriented direction of tips of taper-ratchets, because asymmetric retention forces are formed in the contrasting oriented directions. Balu et al.[80] manipulated the critical sliding angle of a water drop by changing the curvature of the ink pattern. For instance, a semicircular ink pattern allowed a water drop to slide easily in the convex direction, but not in the concave direction. For a straight line pattern printed on SH paper, the drop could move along the line more easily than detach in a direction perpendicular to the line. These studies had demonstrated directional mobility and tunability in the adhesion of water drops on SH surfaces, which was achieved by imposing geometrical constraints on the contact lines by printing specific ink patterns. In another study, according to the microstructure of a rice leaf, Feng et al.[81] prepared the rice-like aligned carbon nanotubes film. The ordered pattern of these films was in two dimensions with different spacing. The water droplets rolled readily across the larger spacing. In the case of slanted pillars generated by e-beam irradiation, Kim et al.[72] reported that the shape and the extent of stooping of the nanohairs could be controlled by the time and power of the e-beam used to irradiate the vertical nanopillars. The stooped nanohairs had a unidirectional feature of frictional adhesion with a remarkably high adhesion force, 11 N/cm^2 in the forward direction and 2.2 N/cm^2 in the reverse direction. As a consequence, it could be attached strongly and detached easily and rapidly. In addition, the adhesion force was maintained over 100 cycles without degradation. Another striking finding was that a simple coating of Pt on the nanohairs avoided the hairs from collapsing and clumping after repeated use. The same group reported a shear attachment of 31 N/cm^2 in the forward direction and 4.1 N/cm^2 in the reverse direction, by combining these slanted pillars with oblique metal deposition.[75] Pressing and shearing an array of vertically aligned CNTs against a surface caused the CNTs to align with the shearing direction and created an anisotropic adhesive with a strong shear adhesion force (100 N/cm^2) and a much lower force (10 N/cm^2).[76] Besides the use of angled fibers, adding spatulas to the tips of microfabricated features could be induced anisotropic adhesion. Murphy et al.[82] developed a fabrication method of bio-inspired microscale fibrillar adhesives where polymer microposts are dipped in uncured PU and then pressed against a counter surface to create the mushroom shapes. These structures exhibited similar shear adhesive strength to gecko lizard feet on smooth surfaces and macroscale support of significant loads gave shearing forces of 100 kPa in the forward direction and only 20 kPa in the reverse direction. The tip angle was identified as a design parameter to control the anisotropy ratio, where higher tip angles facilitated self-releasing behavior when loaded in the releasing direction. They found that the wide flat angled tips of the fibers created a larger contact area and were responsible for the observed shear anisotropy.

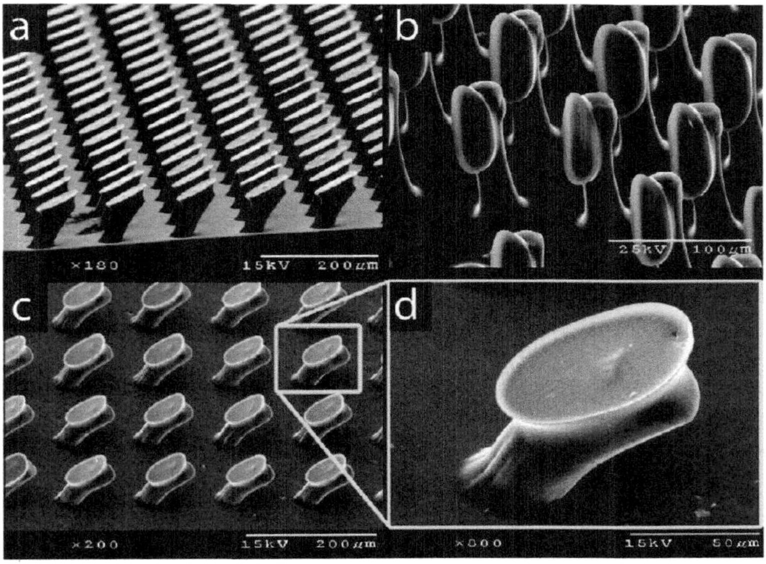

Figure 2.26 Scanning electron microscope images of arrays of 35 mm diameter angled polyurethane microfibers with angled mushroom tips.[82]

Sameoto et al.[83] presented a more deterministic fabrication method utilizing a two-step lithography mold fabrication. They showed that the amount of cap size overhang could control the peel strength of dry adhesives on flat surfaces and a large number of potential geometries could be prototyped rapidly by this technique. Because the exact cap placement, size, and fiber spacing could be modified on a signal mold design, the fabrication of "microtreads" with location specific adhesive properties may have some new macroscale applications. Further efforts have sought to create hierarchically branched structures, which increase the number of contact points and retain sufficient stiffness to prevent mechanical buckling.[84] Jeong et al.[73] presented an hierarchical anisotropic dry adhesive via a two-step UV-assisted replica molding technique. This design not only enabled an adhesive anisotropy due to the inclined features, but also adhesion to rough surfaces due to its compliant hierarchical tips.

Anisotropic adhesives have also been made using SU8, modified soft lithography processes, and laser cutting.[85] These advances in fabrication of anisotropic surfaces and their adhesion properties have been motivated by applications.[82] Jeong et al.[73] demonstrated the application of anisotropic dry adhesives for the handling of thin glass substrates used in liquid crystal displays. A great advantage here is that dry adhesives do not require glue or leave residue, and anisotropy allows easy detachment. Future developments will include active anisotropic structures that can be switched through external stimuli. For instance, dry adhesives using nickel cantilevers that are driven by magnetic fields showed a 40:1 on:off ratio.[86] We may fabricate special surfaces that can switch from nonadhesive to adhesive states on demand and/or where the direction and amount of anisotropy can be adjusted dynamically. We expect additional researches on anisotropic adhesives will occur with the advancement of nanotechnology and anticipate future progress due to extensive application including fluidic devices, anti-fouling, and biomedical coatings.

2.3.3 Directional Driving

Theoretical descriptions of drop motion along a solid surface caused by a wettability gradient are given by Greenspan[87] and Brochard.[88] Surely, anisotropic wetting could be used to drive the droplet. As reported by Kim et al.,[52] controlling fluidic motion via unidirectional wetting on stooped nanohairs was useful for liquid handling in microfluidic devices or flow valves. They reported the fluid velocity along the stooped direction of the polymer nanohairs is six times larger than that in the reverse direction. Similarly, Malvadkar et al.[89] recently used a pin-release droplet ratchet mechanism upon a nanofilm to achieve a unique unidirectional wetting approach and such wetting control held potential for applications such as, microfluidic devices, drag reduction for anti-fouling surfaces, and lab-on-a-chip devices. Zheng et al.[17] reported a novel way to drive directional movement of liquid droplets on a microbeam with a varying or gradient stiffness. According to the energy method they got, the system tends to have the minimum potential energy when the droplet moved to the softer end of the beam. Therefore, a gradient change of the bending stiffness may be utilized to help the directional motion of droplets. Sekeroglu et al.[90] demonstrated the actuation of microgels encapsulated in droplets using a unidirectional nanotextured surface, which moved droplets with low vibration amplitudes by a ratcheting mechanism. The nanofilm carries droplets along the ratchets with minimal drop shape deformation to move the encapsulated soft cargo. Sheng et al.[91] observed directional movements of droplets on the ratchet-like SH surfaces. The observation showed the caterpillar-like crawl of a drop on the inclined SH surfaces as it rolled along the ridge of ratchet. In contrast, along the opposite direction, the movement of the drop only depended on the end of triple phase contact line while the front of contact line was pinned. The sliding angle measurements indicated that the ratchet-like SH surfaces had directional drop retention traits. Moreover, the reduction of the rise angle, the height of the ratchet's ridge, and the volume of the drop can greatly enhance the directional difference of drop retention on the ratchet-like SH surfaces. Therefore, it was concluded that the superhydrophobicity and the periodic ratchet-like microstructures were the keys to the directional drop sliding at 1D level. Xu et al.[92] demonstrated directional transportation of a ball-like water droplet on an SH surface. An SH surface with a water contact angle of as high as 160° and a sliding angle close to 0° was prepared by a sol–gel method. A designed path line was then created on the prepared SH surface. The static CA was measured to be 157° and 144° at the directions parallel and perpendicular to the line, respectively. As a result of this anisotropy, the rolling path of a water droplet can be controlled by the created line. By adjusting the tilting angle of the patterned substrate and the size of the water droplet, it is possible to direct the transportation of a water droplet on an SH surface by path designing.

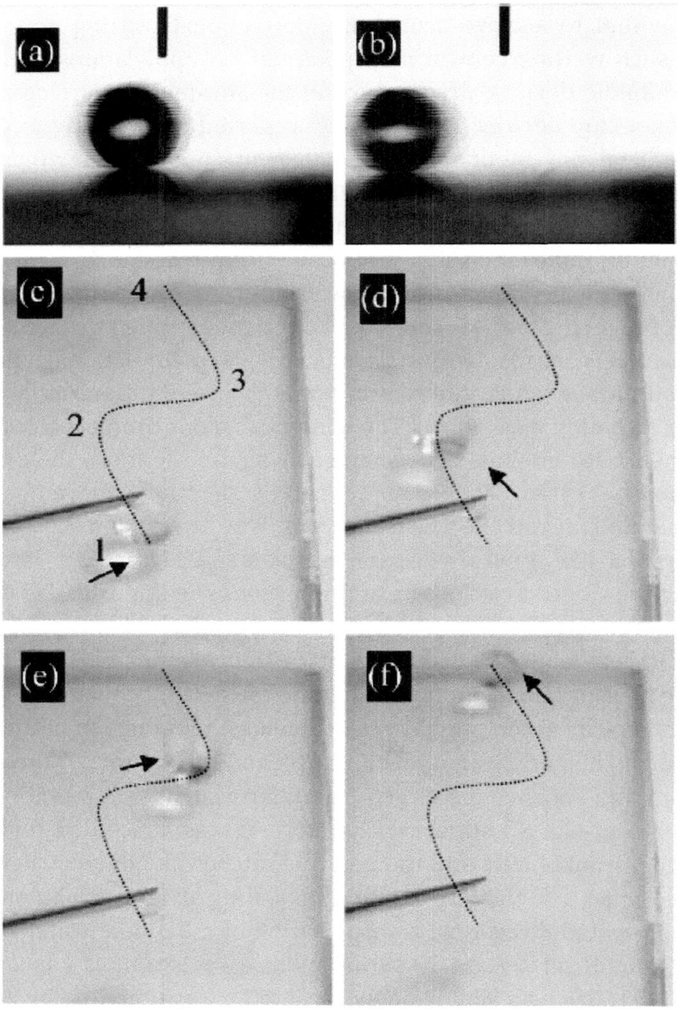

Figure 2.27 Rolling processes of a water droplet (arrowed) (a and b) on an unpatented SH surface and (c–f) on a patterned surface. Note that the rolling path followed the S-shaped line from location 1 to 4.[92]

In 2006, Linke et al.[93] found a special phenomenon of liquids in which they perform self-propelled motion when they were placed in contact with hot surfaces with asymmetric topology. The pumping effect was observed when the liquid was in the Leidenfrost regime for many liquids and over a wide temperature range. They proposed that liquid motion was driven by a viscous force exerted by vapor flow between the solid and the liquid. In 2011, Dupeux et al.[94] found sublimating platelets did the same, and they discussed a possible viscous mechanism for these motions. They thought that the flow of vapor below the levitating material was rectified by the asymmetric teeth of the ratchet, in the direction of descending slopes along each tooth. Hence, the resulting viscous stress can entrain the material in the same direction.

Wu et al.[39] reported that when stimulated by sound wave vibrations, a poly(ethylene terephthalate) (PET) sheet with an aligned oblique nanowire arrays (AONWA) can be unidirectionally moved horizontally, and it could even climb up a 3.6° slope. Similarly, using this kind of PET AONWA, a PS microsphere could be directionally transported for a distance of 0.28 cm up to a 1.6° slope. Feng et al.[95] use wet chemical oxidation followed by fluorization treatment to fabricate a series of SH copper surfaces with nanoribbon structures. Then they systematically studied the influence of surface roughness and the chemical properties of as-prepared surfaces on the spontaneous motion of condensate drops. More obvious spontaneous condensate drop motion was observed on surfaces with a higher extent of fluorization and nanostructures possessing sufficiently narrow spacing and higher perpendicularity. That was due to the stable Cassie state of condensate drops in the dynamic dropwise condensation process that is determined by the nanoscale morphology and local surface energy. Zhang et al.[96] fabricated a series of surfaces with microscale checkerboard patterns consisting of continuous central lines and discontinuous lateral lines and investigated their wetting properties. It was found that the wetting was anisotropic as a result of the anisotropic surface topologies and both central continuous lines and lateral short lines of the checkerboard patterns influenced the surface wetting properties. The drop was easiest and hardest to move in the directions parallel and perpendicular to the central lines, respectively. The water drop moving trajectories are found to deviate from the tilting direction (Fig. 2.28). They were mainly inclined toward the direction of the central lines of the checkerboard pattern and were also influenced by the pointing directions of the free ends of the lateral short lines.

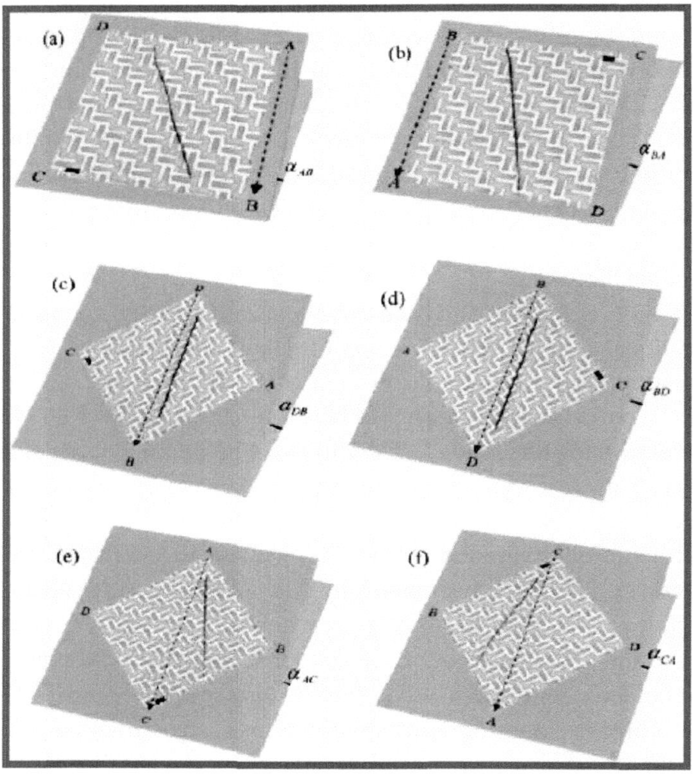

Figure 2.28 Representative motion trajectories of water droplets on the tilted checkerboard surfaces in different tilting directions.

There is another type of anisotropic liquid transportation. This type is directional liquid transportation across the transverse thickness of porous membrane materials, such as cell membranes, nanochannels, and synthetic fabrics. Using a simple membrane model composed of spaced microcylinders. Tian et al.[56] showed unambiguously that a wettability gradient along the thickness direction could cause anisotropic liquid penetration behavior, which was attributed to the coupled effects of local geometrical angle and contact angle on the critical breakthrough pressure. The anisotropic ratio of critical breakthrough pressure could be improved by reducing the spacing ratio of the cylinder membrane as well as by increasing the wettability gradient (Fig. 2.29).

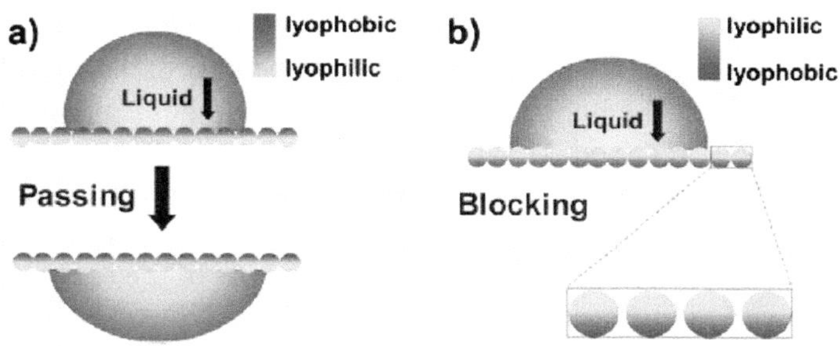

Figure 2.29 Anisotropic liquid penetration through a microcylinder membrane with a wettability gradient along its thickness.

2.3.4 Other Applications

Besides above, there are many other important applications. For example, Kannan et al.[97] reported an experimental investigation of water drops impacting onto solid surfaces with anisotropic wetting. It was observed that the beating phenomenon of impacting drop parameters was caused by the frequency difference between the dynamic contact angle oscillations of impacting drop liquid oriented perpendicular and parallel to the direction of grooves on the grooved surface. The beat frequency was almost independent of the impact drop Weber number. Further experimental measurements showed that the grooved surface with larger wetting anisotropy may be expected to show a dominant beating phenomenon. The phenomenon was gradually damped out with time and was fully unrecognizable at higher drop impact Weber numbers. Malouin et al.[98] showed that droplet rebounding on textured surfaces can be significantly influenced by the uniformity of the surface roughness. By engineering nonuniform textures, that is, roughness gradients on the surface, they were able to not only manipulate the axial rebound of the droplet, but also introduced a prescribed lateral component to the rebound trajectory. The presence of a surface roughness gradient set up a wetting gradient, which in turn introduced a prescribed lateral component to the trajectory of a rebounding droplet that was not observed in cases with uniform roughness. The use of surfaces with nonuniform textures may then enable practitioners to control the placement and trajectory of droplets after impact with a surface. Inspired by insect water strider, Liu et al.[6] designed a model device named "oil strider" that could float stably at the oil/water interface without contamination by oil. The floating capability of the oil strider originated from the huge underwater superoleophobic supporting force its "legs" received. They prepared the micro/nanohierarchical structured copper-oxide-coated copper wires, acting as the artificial legs of oil strider, by a simple base-corrosion process. The surface structures and hydrophilic chemical components of the coatings on copper wires induced the huge superoleophobic force at the oil/water interface, to support the oil strider from sinking into the oil.

In summary, the typical potential applications of anisotropic wetting include flow valves, microfluidics, biosensors, evaporation-driven pattern formation, droplet transport, microreactors, and tunable/switchable surfaces. Notably, we have not attempted to cover all potential applications of anisotropic wetting. With the development of science and technology, more and more fields would benefit from anisotropic wetting.

2.4 Outlook

The rapid development of research about anisotropic wetting offers a variety of exciting opportunities for further exploration. Some substantial progress has been made in fabricating surfaces exhibiting anisotropic wetting behavior in recent years, owing to recent development in micro-/nanofabrication, and material synthesis. However, compared to research on SH wetting, significantly research of anisotropic wetting is needed. For instance, additional insight and research is needed for optimizing and scaling up fabrication methods to achieve low cost, mass scaling, and biocompatibility. And for many industrial applications, scalable fabrication on metal and ceramic surfaces, continued progress in understanding anisotropic wetting behavior and fabricating technology will need to develop. Obviously, there are great opportunities and challenges for further developments, to achieve an eventual goal to enable design and fabrication of heterogeneous surface features over large areas. And, some standards are needed to create to control of complex surface textures, which is especially important for manufacturing. Therefore, the most impressive advances in this field would combine basic understanding of interface properties with novel fabrication principles that can be adapted to large-scale production.

In addition, according to the mention above, there have been some initial investigations on the impact of droplets on anisotropic wetting surfaces. The vapor droplet condensation on anisotropic wetting surfaces is an interesting field. However, less research exists on the dynamic processing of evaporation of droplets on anisotropic wetting surface. With combination of prepatterned surfaces and anisotropic wetting surfaces, complex droplets could be formed to guide fluidic flow and control the droplet shapes, which is useful for droplet fluidic devices. Numerical modeling and simulations are effective for predicting anisotropic wetting behavior on surfaces. Recently, most simulations involve topographic patterned surfaces or SH surfaces. Less modeling research currently involves chemically patterned surfaces with anisotropic wetting characteristics. Considerable work is required to develop more suitable models and simulations aimed at fitting experimental results.

We hope that advancements in anisotropic wetting research will assist in development of devices with multifunctional and controllable utilities. In the case of multifunctional devices, the combination of anisotropic surface wetting properties with other properties may produce a device with desired functions. For instance, anisotropy of both adhesion and wetting properties was observed on directional structures, thereby providing multifunctional abilities.[77] Anisotropic wetting also offers a platform for fabricating a controllable device. We could add an extra stimulus (e.g., optical, electrical, heat, magnetic) to achieve the smart response. It is also possible to manipulate the fluids in micro- and nanoscale on anisotropic wetting surface with simultaneous employment of light and electrical field. In the near future, we expect that advancements in anisotropic wetting research will be exploited to fabricate smart devices having multifunctional properties and/or tunable properties with extra controlling forces. The change may occur between conditions such as anisotropic/isotropic, hydrophobic/hydrophilic, or static stable/dynamic sliding with potential importance for biomedical devices. In addition, combination of topographical anisotropy and gradients in surface chemistry could lead to highest drop shape anisotropy and directional spreading of water droplet. We believe some progress will be made in near future for anisotropic wetting on gradient surfaces, developed to be combined with chemical gradient surface.

We have appreciated the current, rapid progress in anisotropic wetting research and believe research will produce new findings applicable for various advancements, such as, fluidic devices, sensors, microreactors, biomedical devices, self-cleaning. Clearly, the applications of anisotropic wetting hold potential to benefit human health, energy, and the environment in the further.

References

1. Yoshioka S, Kinoshita S (2004) Wavelength–selective and anisotropic light–diffusing scale on the wing of the Morpho butterfly, *Proc R Soc Lond B Biol Sci*, 271, 581–587.
2. Wagner T, Neinhuis C, Barthlott W (1996) Wettability and contaminability of insect wings as a function of their surface sculptures, *Acta Zool*, 77, 213–225.
3. Zheng Y, Gao X, Jiang L (2007) Directional adhesion of superhydrophobic butterfly wings, *Soft Matter*, 3(2), 178–182.
4. Liu K, Jiang L (2011) Bio-inspired design of multiscale structures for function integration, *Nano Today*, 6(2), 155–175.
5. Kim YH, et al. (2009) Fabrication of a superhydrophobic surface from a smectic liquid-crystal defect array, *Adv Funct Mater*, 19(18), 3008–3013.
6. Liu X, et al. (2012) Bioinspired oil strider floating at the oil/water interface supported by huge superoleophobic force, *ACS Nano*, 6(6), 5614–5620.
7. Guo P, Zheng Y, Liu C, Ju J, Jiang L (2012) Directional shedding-off of water on natural/bio-mimetic taper-ratchet array surfaces, *Soft Matter*, 8(6), 1770–1775.
8. Bossart O, Calzaferri G (2008) Self assembly of zeolite L crystals on biological self-cleaning surfaces, *Microporous Mesoporous Mater*, 109(1–3), 392–397.

9. Xi J, Jiang L (2008) Biomimic superhydrophobic surface with high adhesive forces, *Ind Eng Chem Res,* 47(17), 6354–6357.

10. Mei H, et al. (2011) Multi-level micro-/nanostructures of butterfly wings adapt at low temperature to water repellency, *Soft Matter,* 7(22), 10569–10573.

11. Wanasekara ND, Chalivendra VB (2011) Role of surface roughness on wettability and coefficient of restitution in butterfly wings, *Soft Matter,* 7(2), 373–379.

12. Sun G, Fang Y, Cong Q, Ren L-q (2009) Anisotropism of the non-smooth surface of butterfly wing, *J Bionic Eng,* 6(1), 71–76.

13. Xiu Y, Zhu L, Hess DW, Wong CP (2007) Hierarchical silicon etched structures for controlled hydrophobicity/superhydrophobicity, *Nano Lett,* 7(11), 3388–3393.

14. Shastry A, Case MJ, Böhringer KF (2006) Directing droplets using microstructured *surfaces, Langmuir,* 22, 6161–6167.

15. Mishchenko L, et al. (2010) Design of ice-free nanostructured surfaces based on repulsion of impacting water droplets, *ACS Nano,* 4(12), 7699–7707.

16. Wu D, et al. (2011) Three-level biomimetic rice-leaf surfaces with controllable anisotropic sliding, *Adv Funct Mater,* 21(15), 2927–2932.

17. Zheng X-P, et al. (2008) Elasticity-driven droplet movement on a microbeam with gradient stiffness: A biomimetic self-propelling mechanism, *J Colloid Inter Sci,* 323(1), 133–140.

18. Wenzel RN (1936) Resistance of solid surfaces to wetting by water, *Ind Eng Chem,* 28(8), 988–994.

19. Cassie ABD, Baxter S (1944) Wettability of porous surfaces, *Trans Faraday Soc,* 40(0), 546–551.

20. Lafuma A, Quéré D (2003) Superhydrophobic states, *Nat Mater,* 2(7), 457–460.

21. Shirtcliffe NJ, McHale G, Newton MI, Chabrol G, Perry CC (2004) Dual-scale roughness produces unusually water-repellent surfaces, *Adv Mater,* 16(21), 1929–1932.

22. Karmouch R, Ross GG (2010) Experimental study on the evolution of contact angles with temperature near the freezing point, *J Phys Chem C,* 114(9), 4063–4066.

23. Zaidman B (1989) Use of the dimensional relation approach in a computerized model for the simulation of a batch production plant, *Ind Eng Chem Res,* 28(7), 998–1003.

24. He M, Li H, Wang J, Song Y (2011) Superhydrophobic surface at low surface temperature, *Appl Phys Lett,* 98(9), 093118.

25. Zheng Y, et al. (2010) Directional water collection on wetted spider silk, *Nature,* 463(7281), 640–643.

26. Hou Y, Chen Y, Xue Y, Zheng Y, Jiang L (2012) Water collection behavior and hanging ability of bioinspired fiber, *Langmuir,* 28(10), 4737–4743.

27. Xia D, Johnson LM, López GP (2012) Anisotropic wetting surfaces with one-dimesional and directional structures: Fabrication approaches, wetting properties and potential applications, *Adv Mater,* 24(10), 1287–1302.

28. Bliznyuk O, Vereshchagina E, Kooij ES, Poelsema B (2009) Scaling of anisotropic droplet shapes on chemically stripe-patterned surfaces, *Phys Rev E,* 79(4), 041601.

29. Sommers AD, Jacobi AM (2006) Creating micro-scale surface topology to achieve anisotropic wettability on an aluminum surface, *J Micromech Microen,* 16(8), 1571.

30. Xia D, Brueck SRJ (2008) Strongly anisotropic wetting on one-dimensional nanopatterned surfaces, *Nano Lett,* 8(9), 2819–2824.

31. Wu S-Z, et al. (2010) One-step preparation of regular micropearl arrays for two-direction controllable anisotropic wetting, *Langmuir,* 26(14), 12012–12016.

32. Wu D, et al. (2010) A simple strategy to realize biomimetic surfaces with controlled anisotropic wetting, *Appl Phys Lett,* 96(5), 053704.

33. Sato H, Matsumura H, Keino S, Shoji S (2006) An all SU-8 microfluidic chip with built-in 3D fine microstructures, *J Micromech Microeng,* 16(11), 2318.

34. Zhang D, et al. (2010) Wetting characteristics on hierarchical structures patterned by a femtosecond laser, *J Micromech Microeng,* 20(7), 075029.

35. Chen F, et al. (2010) Anisotropic wetting on microstrips surface fabricated by femtosecond laser, *Langmuir,* 27(1), 359–365.

36. Gao X, Yao X, Jiang L (2007) Effects of rugged nanoprotrusions on the surface hydrophobicity and water adhesion of anisotropic micropatterns, *Langmuir,* 23(9), 4886–4891.

37. Wang T, et al. (2010) Elliptical silicon arrays with anisotropic optical and wetting properties, *Langmuir,* 26(16), 13715–13721.

38. Zhang X, et al. (2009) Morphology and wettability control of silicon cone arrays using colloidal lithography, *Langmuir,* 25(13), 7375–7382.

39. Wu W, Cheng L, Bai S, Wang ZL, Qin Y (2012) Directional transport of polymer sheet and a microsphere by a rationally aligned nanowire array, *Adv Mater*, 24(6), 817–821.

40. Neuhaus S, Spencer ND, Padeste C (2011) Anisotropic wetting of microstructured surfaces as a function of surface chemistry, *ACS Appl* Mater Interface 4(1), 123–130.

41. Liu L, Jacobi AM, Chvedov D (2009) A surface embossing technique to create micro-grooves on an aluminum fin stock for drainage enhancement, *J Micromech Microeng*, 19(3), 035026.

42. Yang J, Rose FRAJ, Gadegaard N, Alexander MR (2009) Effect of sessile drop volume on the wetting anisotropy observed on grooved surfaces, *Langmuir*, 25(5), 2567–2571.

43. Zhang T, et al. (2011) Bio-inspired anisotropic micro/nano-surface from a natural stamp: Grasshopper wings, *Soft Matter*, 7(18), 7973–7975.

44. Mele E, Girardo S, Pisignano D (2012) Strelitzia reginae leaf as a natural template for anisotropic wetting and superhydrophobicity, *Langmuir*, 28(11), 5312–5317.

45. Zhao W, Wang L, Xue Q (2010) Fabrication of low and high adhesion hydrophobic Au surfaces with micro/nano-biomimetic structures, *J Phys Chem C*, 114(26), 11509–11514.

46. Huang JY, Wang XD, Wang ZL (2006) Controlled replication of butterfly wings for achieving tunable photonic properties, *Nano Lett*, 6(10), 2325–2331.

47. Pang C, Kang D, Kim T-i, Suh K-Y (2011) Analysis of preload-dependent reversible mechanical interlocking using beetle-inspired wing locking device, *Langmuir*, 28(4), 2181–2186.

48. Tawfick S, Hart AJ, De Volder M (2012) Capillary bending of Janus carbon nanotube micropillars, *Nanoscale*, 4(13), 3852–3856.

49. Zhang F, Low HY (2007) Anisotropic wettability on imprinted hierarchical structures, *Langmuir*, 23(14), 7793–7798.

50. Khare K, Zhou J, Yang S (2009) Tunable open-channel microfluidics on soft Poly(dimethylsiloxane) (PDMS) substrates with sinusoidal grooves, *Langmuir*, 25(21), 12794–12799.

51. Zhang Z, Zhang T, Zhang YW, Kim K-S, Gao H (2011) Strain-controlled switching of hierarchically wrinkled surfaces between superhydrophobicity and superhydrophilicity, *Langmuir*, 28(5), 2753–2760.

52. Kim T-i, Suh KY (2009) Unidirectional wetting and spreading on stooped polymer nanohairs, *Soft Matter,* 5(21), 4131–4135.
53. Kwak MK, Jeong H-E, Kim T-i, Yoon H, Suh KY (2010) Bio-inspired slanted polymer nanohairs for anisotropic wetting and directional dry adhesion, *Soft Matter,* 6(9), 1849–1857.
54. Balu B, Berry AD, Hess DW, Breedveld V (2009) Patterning of superhydrophobic paper to control the mobility of micro-liter drops for two-dimensional lab-on-paper applications, *Lab Chip,* 9(21), 3066–3075.
55. Prakash M, Bush JWM (2011) Interfacial propulsion by directional adhesion, *Int J Non-Lin,* Mech 46(4), 607–615.
56. Tian X, Li J, Wang X (2012) Anisotropic liquid penetration arising from a cross-sectional wettability gradient, *Soft Matter,* 8(9), 2633–2637.
57. Drelich J, Wilbur JL, Miller JD, Whitesides GM (1996) Contact angles for liquid drops at a model heterogeneous surface consisting of alternating and parallel hydrophobic/hydrophilic strips, *Langmuir,* 12(7), 1913–1922.
58. Bauer J, et al. (2000) The major protein import receptor of plastids is essential for chloroplast biogenesis, *Nature,* 403(6766), 203–207.
59. Sharma R, Strano MS (2009) Centerline placement and alignment of anisotropic nanotubes in high aspect ratio cylindrical droplets of nanometer diameter, *Adv Mater,* 21(1), 60–65.
60. Jokinen V, Sainiemi L, Franssila S (2008) Complex droplets on chemically modified Silicon nanograss, *Adv Mater,* 20(18), 3453–3456.
61. Chu K-H, Xiao R, Wang EN (2010) Uni-directional liquid spreading on asymmetric nanostructured surfaces, *Nat Mater,* 9(5), 413–417.
62. Autumn K, et al. (2000) Adhesive force of a single gecko foot-hair, *Nature,* 405(6787), 681–685.
63. Arzt E, Gorb S, Spolenak R (2003) From micro to nano contacts in biological attachment devices, *Proc Nat Am Soc,* 100(19), 10603–10606.
64. Parness A, et al. (2009) A microfabricated wedge-shaped adhesive array displaying gecko-like dynamic adhesion, directionality and long lifetime, *J R Soc Interface,* 6(41), 1223–1232.
65. Huber G, et al. (2005) Evidence for capillarity contributions to gecko adhesion from single spatula nanomechanical measurements, *Proc Nat Am Soc,* 102(45), 16293–16296.
66. Lee DY, Lee DH, Lee SG, Cho K (2012) Hierarchical gecko-inspired nanohairs with a high aspect ratio induced by nanoyielding, *Soft Matter,* 8(18), 4905–4910.

67. Murphy MP, Kim S, Sitti M (2009) Enhanced adhesion by gecko-inspired hierarchical fibrillar adhesives, *ACS Appl Mater Interfaces,* 1(4), 849–855.

68. Scherge M, Gorb SN (2000) Microtribology of biological materials (translated from English), *Tribol Lett,* 8(1), 1–7 (in English).

69. Yao H, Gao H (2006) Mechanics of robust and releasable adhesion in biology: Bottom–up designed hierarchical structures of gecko, *J Mech Phys Solids,* 54(6), 1120–1146.

70. Niederegger S, Gorb S (2006) Friction and adhesion in the tarsal and metatarsal scopulae of spiders (translated from English), *J Comp Physiol A,* 192(11), 1223–1232 (in English).

71. Jeong HE, Lee SH, Kim P, Suh KY (2006) Stretched polymer nanohairs by nanodrawing, *Nano Lett,* 6(7), 1508–1513.

72. Kim T-i, Jeong HE, Suh KY, Lee HH (2009) Stooped nanohairs: Geometry-controllable, unidirectional, reversible, and robust gecko-like dry adhesive, *Adv Mater,* 21(22), 2276–2281.

73. Jeong HE, Lee J-K, Kim HN, Moon SH, Suh KY (2009) A nontransferring dry adhesive with hierarchical polymer nanohairs, *PNAS,* 106(14), 5639–5644.

74. Mekaru H, Takahashi M (2009) Frequency and amplitude dependences of molding accuracy in ultrasonic nanoimprint technology, *J Micromech Microeng,* 19(12), 125026.

75. Yoon H, et al. (2009) Adhesion hysteresis of Janus nanopillars fabricated by nanomolding and oblique metal deposition, *Nano Today,* 4(5), 385–392.

76. Qu L, Dai L, Stone M, Xia Z, Wang ZL (2008) Carbon nanotube arrays with strong shear binding-on and easy normal lifting-off, *Science,* 322(5899), 238–242.

77. Zhang J, Cheng Z, Zheng Y, Jiang L (2009) Ratchet-induced anisotropic behavior of superparamagnetic microdroplet, *Appl Phys Lett,* 94, 144104.

78. Pang C, et al. (2012) Bioinspired reversible interlocker using regularly arrayed high aspect-ratio polymer fibers, *Adv Mater,* 24(4), 475–479.

79. Pang C, Kim SM, Rahmawan Y, Suh K-Y (2012) Beetle-inspired bidirectional, asymmetric interlocking using geometry-tunable nanohairs, *ACS Appl Mater Interfaces,* 4(8), 4225–4230.

80. Balu B, Berry AD, Patel KT, Breedveld V, Hess DW (2011) Directional mobility and adhesion of water drops on patterned superhydrophobic surfaces, *J Adhes Sci Technol,* 25(6–7), 627–642.

81. Feng L, et al. (2002) Super-hydrophobic surfaces: From natural to artificial, *Adv Mater,* 14(24), 1857–1860.
82. Murphy MP, Aksak B, Sitti M (2009) Gecko-inspired directional and controllable adhesion, *Small,* 5(2), 170–175.
83. Sameoto D, Menon C (2009) Direct molding of dry adhesives with anisotropic peel strength using an offset lift-off photoresist mold, *J Micromech Microeng,* 19(11), 115026.
84. Jeong HE, Suh KY (2009) Nanohairs and nanotubes: Efficient structural elements for gecko-inspired artificial dry adhesives, *Nano Today,* 4(4), 335–346.
85. Greiner C, Arzt E, del Campo A (2009) Hierarchical gecko-like adhesives, *Adv Mater,* 21(4), 479–482.
86. Northen MT, Greiner C, Arzt E, Turner KL (2008) A gecko-inspired reversible adhesive, *Adv Mater,* 20(20), 3905–3909.
87. Greenspan HP (1978) On the motion of a small viscous droplet that wets a surface, *J Fluid Mech,* 84(01), 125–143.
88. Brochard F (1989) Motions of droplets on solid surfaces induced by chemical or thermal gradients, *Langmuir,* 5(2), 432–438.
89. Malvadkar NA, Hancock MJ, Sekeroglu K, Dressick WJ, Demirel MC (2010) An engineered anisotropic nanofilm with unidirectional wetting properties, *Nat Mater,* 9(12), 1023–1028.
90. Sekeroglu K, Gurkan UA, Demirci U, Demirel MC (2011) Transport of a soft cargo on a nanoscale ratchet, *Appl Phys Lett,* 99(6), 063703.
91. Sheng X, Zhang J (2011) Directional motion of water drop on ratchet-like superhydrophobic surfaces, *Appl Surf Sci,* 257(15), 6811–6816.
92. Xu QF, Wang JN, Smith IH, Sanderson KD (2008) Directing the transportation of a water droplet on a patterned superhydrophobic surface, *Appl Phys Lett,* 93(23), 233112.
93. Linke H, et al. (2006) Self-propelled leidenfrost droplets, *Phys Rev Lett,* 96(15), 154502.
94. Dupeux G, et al. (2011) Viscous mechanism for Leidenfrost propulsion on a ratchet, *EPL,* 96(5), 58001.
95. Feng J, Qin Z, Yao S (2012) Factors affecting the spontaneous motion of condensate drops on superhydrophobic copper surfaces, *Langmuir,* 28(14), 6067–6075.
96. Zhang X, Cai Y, Mi Y (2011) Anisotropic wetting on checkerboard-patterned surfaces, *Langmuir,* 27(15), 9630–9637.

97. Kannan R, Vaikuntanathan V, Sivakumar D (2011) Dynamic contact angle beating from drops impacting onto solid surfaces exhibiting anisotropic wetting, *Colloid Surfaces A,* 386(1–3), 36–44.

98. Bernard A, Malouin J, Koratkar NA, Hirsa AH, Wang Z (2010) Directed rebounding of droplets by microscale surface roughness gradients, *Appl Phys Lett,* 96(23), 234103.

Chapter 3

Spider Silk Effect: Gradient Micro- and Nanostructures

Having been evolving for billions of years, many creatures in the nature have various unusual abilities to survive in tough conditions. Bio-inspired by nature, designing and fabricating smart functional materials have become a very promising field of research. Wettability is the one of the most important subdisciplines of this field to prepare bio-inspired smart materials. Some biological surfaces show special wettability features such as lotus, water strider,[1] rose, butterfly, spider silk, rice leaf, and desert beetle. For example, the capture silk of the *cribellate* spider (*Uloborus walckenaerius*) has a water-collecting ability resulting from a unique fiber structure that is formed after wetting with the "wet-rebuilt" fibers characterized by periodic spindle knots made of random nanofibrils and separated by joints made of aligned nanofibrils. These structural features result in a surface energy gradient between the spindle knots and the joints and also in a difference in Laplace pressure, with both factors acting together to achieve continuous condensation and directional collection of water drops around spindle knots. We named it as "Effect of Spider Silk." Submillimeter-sized liquid drops have been driven

Bio-Inspired Wettability Surfaces: Developments in Micro- and Nanostructures
Yongmei Zheng

ISBN 978-981-4463-60-7 (Hardcover), 978-981-4463-61-4 (eBook)
www.panstanford.com

by surface energy gradients or a difference in Laplace pressure. However, until now only one force is difficult to overcome the larger hysteresis effects to drive a micrometer-sized drops movement. By tapping into both driving forces, spider silk can achieve this task. Inspired by spider silk, Zheng's team designed artificial fibers to mimic the structural features of silk and exhibit its directional water collecting ability.

In this chapter, we give a brief review about the effect of spider silks and the functional fibers with unique wettability inspired by spider silks. The spider's web can pin some shiny water drops. This beautiful sight gives us an inspiration to mimic the spider silks to further design functional fiber materials. We introduce some methods to develop the process of fabrication, such as dip coating, fluid coating, and electrodynamic coating. Figure 3.1 shows the summarization about the designing strategies and applications of bio-inspired functional fibers with unique wettability. Designing both structural features and chemical components of the fiber surface can obtain novel functional fibers. For example, Fig. 3.1d shows some function designs by using some light, pH, or thermal stimuli molecules modified on the fiber surfaces to control the moving direction of water droplets. Or, designing asymmetric structures or wettability gradient on the spindle knot achieves the unidirectional motion of water droplets (Figs. 3.1a and c). This would achieve fast and high-efficiency liquid transportation along the fiber. In addition, we look out for bio-inspired functional fibers with unique wettability in many fields, such as water collection (Fig. 3.1d), smart catalysis (Fig. 3.1e), filtration, and sensing (Fig. 3.1f). Figure 3.1d shows the water-collecting efficiency can be increased by directionally transporting small water droplets together to coalesce into large droplets. Smart catalysis can expedite the transporting speed of water drop toward or away from the spindle knot, shown in Fig. 3.1e. Figure 3.1f shows that the tiny liquid drops on the bio-inspired functional fiber can capture small amount of materials in the air (e.g., particles). Then by directionally collecting the liquid droplets toward the spindle knot further concentrates or filtrates the small materials.

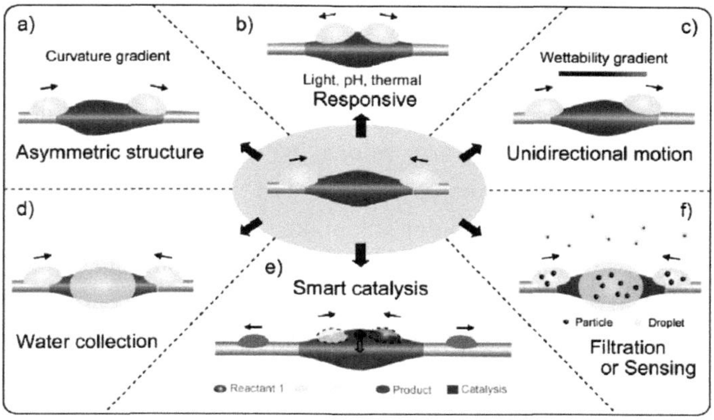

Figure 3.1 Schematic illustration of designing concepts and applications of bio-inspired functional fibers with unique wettability.

Bio-inspired by the water collection ability of cribellate spider silk, here, we will introduce in detail the structure, mechanical properties, elasticity, and water sensitivity of spider silk. We further reveal the relationship of its unique wettability with its elaborate micro- and nanostructures. Series researches of fabricating methods and water collection ability of artificial spider silk will be shown in this chapter. At the same time, we will introduce some applications about this new material.

3.1 Spider Silk

After millions of years of evolution, many animals are able to create a filamentous fiber material. The use of the filamentous fiber material also has thousands of years of history, including fabrics, medical wound dressings, as well as military applications.[2] Silk is a remarkable biological product. The most natural filamentous fibers derived from silkworm (Bombyx moth) whose food is mulberry leaves. This silkworm is very easy to farm, and its history can be traced back to 5,000 years ago in China. The worm silk is used for textile qualities, and it is easy to obtain by farm the silkworm. However, the spider silk is different. It is difficult to get a large number of fibers. The reason is that the spider is very pugnacious and not easily domesticated. Although spider silks have no commercial market yet, their utilization (primarily the Araneidae web) also has a long history. Their study has led to revolutionary new insights into silk properties because, unlike insect silks, they have evolved to perform best under tension and so are uniform in material qualities along the length of the fiber. The major advantage of the excellent mechanical and biocompatibility of spider silk make it useful as a fishing line or wound covering material. Natural spider silk is a biological material, and the main ingredient is a kind of protein secreted by spider glands. Undoubtedly, it is one of the nature's most excellent structural and functional materials.

3.1.1 Classification of Spider Silk

"Silk," the English word in the Oxford Dictionary, derives from a Baltic/Slavic precursor. And this word was advised as an old name for all type of textile. But now, we define silk primarily as the fiber spun by the larva of the silk. We also collect it by decomposing some insect's cocoon. The "silk" also includes the fiber properties, for example caddisflies, mites, and spiders. We even call the filaments of the ears of corn as "silk." Here, we want to give more details about the biophysical or biochemical similarity with the animal silks.[3]

You can find spider orb web in many parts of the world, such as under the tree, on the leaves. Its radius ranges from a few centimeters to several meters. We discovered a spider in Madagascar recently. This spider can weave a silk of up to 25 m. It is difficult to distinguish the difference between spider silks with our naked eyes. In fact, a spider web is constituted by as many as five different types of spider silk. Figure 3.2 shows that a maternal spider (a kind of orb weaving spider) can weave seven kinds of spider silk, even including a specialized silk that is used as egg cases. Why does the spider produce different kinds of silk? Different silks each with specific properties appear to be optimized to perform key functional roles. In fact, the spider needs to adapt to the environment in the process of evolution. To achieve this purpose, it can magically weave different kinds of silks.

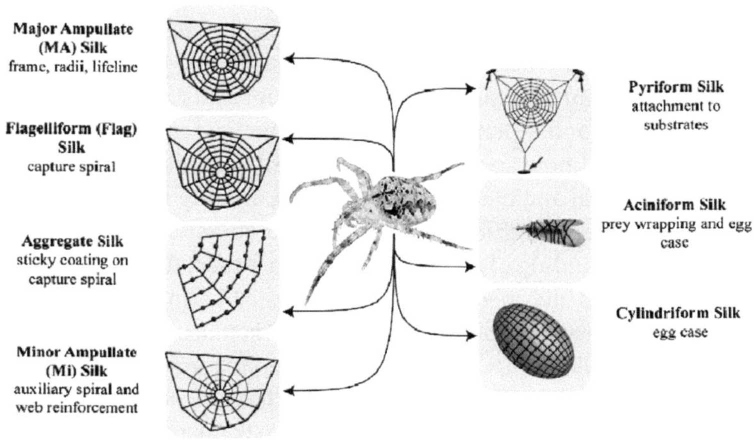

Figure 3.2 Schematic overview of different silk types produced by female orb weaving spiders (Araneae).

Silkworm silk is composed of fiber (bave) or comprises a doublet of fibroin filaments (brins), which is surrounded by several thick layers of sericin coating. Both fibroin and sericin are composed of proteins. But the fibroin and sericin have different composition and properties. The two brins are produced and coated in separate ducts and pressed together; the sericin hardens in the air and on the cocoon to form the conjoining bave and tough cocoon wall composite. Spider silk also consists of two filaments, but they are easily separated. Their coating consists of a varied number of small molecules of nonprotein compounds, which is much thinner, that is, the benchmark of spider silks, the dragline, and frame silk of the orb weaving tribe of araneid spiders, such as the garden orb spider *Araneus diadematus* or the golden silk spider *Nephila clavipes*. Orbweavers like Nephila produce minor-ampullate silk to accompany the major ampullate, as well as flagelliform silk. This can form the core filaments of the orb web's capture thread. Another "silk" is from the aggregate gland to coat the capture thread filaments. This "silk" is not hard but an aqueous solution of small and highly hygroscopic peptides as well as sticky glycoproteins.[3]

Because of different functions, the mechanical properties of spider silk are also very different. The stress ranges from 0.02 to

1.7 GPa (*Caerostris darwini* dragline silk) in terms of the different types of spider silk. And steel stress is 1.2 GPa. The tensile properties of spider silk are also very good, ranging from 10% to 500%. What is more interesting is that most spider silks have both good strength and good tensile properties. So the spider silk has a high toughness. We know some classic animal silks share many traits in their biophysical and biochemical make-up, although there are a fair number of interesting differences. But the spider silk is probably the most outstanding of all biological fibers, especially in its general structure and in its mechanical properties. Dragline spider silk is significantly stronger and more extensible than any other silk, such as silkworm silk. Figure 3.3 shows the Nephila golden silk glands as well as the use for each silk and its mechanical properties in the form of comparative experimental stress–strain plots, where the highlighted line in each graph is that of the graph label; the symbols D and W in the flagelliform graph are dry and wet, respectively.

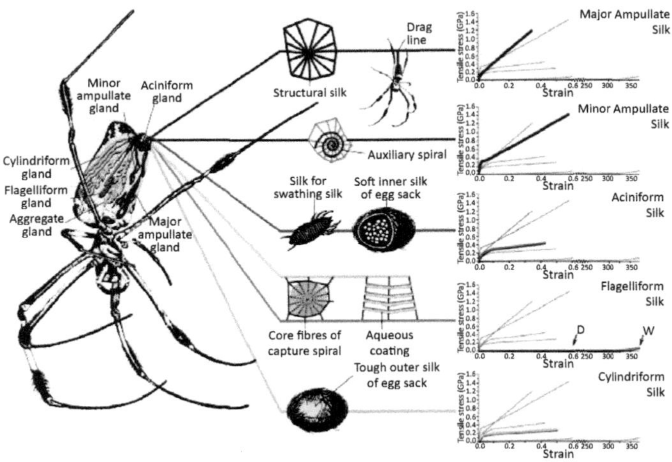

Figure 3.3 Nephila golden silk spider showing its silk glands as well as the use for each silk and its mechanical properties in the form of comparative experimental stress–strain plots, where the highlighted line in each graph is that of the graph label; the symbols D and W in the flagelliform graph are dry and wet, respectively.

Silk has amazing multiplicity of types and uses, and has excellent both mechanical properties and chemical composition. Despite these differences, all spider silks derive from one common type of ancestral silk gland. After nearly 400 million years of evolution, all these glands have a common origin. The selection of different functionalities has led to diversification and to biochemical tuning as well as to mechanical specializations and to temporary tensile adjustments. Due to the diversity of spider silks, it is possible for obtaining one natural properties biomaterial on both the chemical and the physical level. The varieties of spider silks open a window into the structure–function relationships of many different properties.[3]

This gives us another research direction that all of the spider silks have a good viscoelasticity. Spider capture silk is a natural material that outperforms almost any synthetic material in its combination of strength and elasticity. A typical orb-weaving spider is supported by frame silk and dragline silk, and constitute a skeleton with radial spokes silk adding to viscous capture silk to form a mesh structure of the ring. The capture silk is one kind of sticky spiral of orb-weaving spiders. We have investigated the capture spiral threads from Araneus, who is an ecribellate orb-weaving spider. The flagelliform protein composed of molecular and supramolecular structure is the major protein of these threads. And the protein is from amino acid sequences, force spectroscopy and stretching of bulk capture web.[4] Soft annular capture silk and rigid spokes silk together constitute the web, which can absorb the ability of impact of the insect to capture it.[5,6] Figure 3.4 shows the structure of two kinds of capture silks. In view of the spider silk is strong and flexible, people use it to make bulletproof vests, helmets, parachutes, etc. People also want to use it to replace the brittle material in low temperature because it can maintain a good flexibility at low temperatures. In fact, spider silk also has potential applications as function materials in bridge construction, biomedical, and other fields.

With reference to the presence or absence of a cribellum, spider can be classified into two categories as cribellates or ecribellates. The structure of the capture threads of ecribellates is simple, which is composed of main threads and nanofibers covering on the main thread surface. In Araneidae, for instance, they are composed of two axial fibers that are with a series of droplets of viscous glue. The silk captures food by van der Waals force and electrostatic force. The energy of prey impacting web is absorbed by the nanofibers fracture. Accordingly, the capture threads of cribellates are much more complex. It is composed of core threads and sticky droplets on the core threads. And innumerable numbers of thin fibrils are supported by the two axial fibers in one web. Except axial fibers and cribellum fibrils, one more category of fibers has always been observed in the capture threads of cribellates, except for uloboridae. Their number and structure are different according to the families. It was shown that a comparable third component is also present in the capture threads of uloborids. Thus, three different categories of components are typical for the capture threads of cribellate spiders in general.[7] Due to the glue and adhesiveness of extremely thin fibrils (diameter 20–30 nm), the spider can capture preys. When the preys impact spider silk, the curled core threads on the sticky droplets is opened, which absorbs the impact energy like wheels.

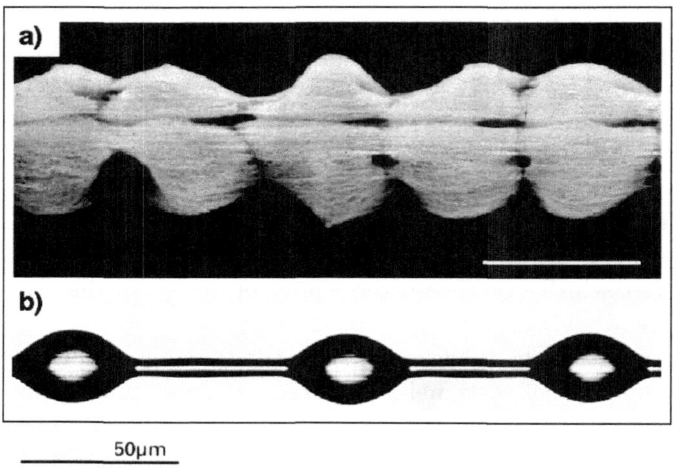

Figure 3.4 Two kinds of spider's capture silk: (a) scanning electron microscopic image of cribellar capture silk with periodic "puff." (b) Optical image of adhesive ecribellar capture silk with flue droplets. Scale bars are 50 μm.

Most of recent work on silk is about the experimental measurements of physical properties including mechanical performance and elasticity, which has been reviewed previously. However, some researchers find that silk protein structure plays an important role on the mechanical properties and elasticity. Some research groups work in this field in order to develop a model to show the complex composition and microstructure of this intriguing material. These new models aim to identify the underlying processes to control chemical–physical properties as the design tools for most natural biomaterials.

Silk typically consists of two types of proteins: one large and one much smaller. The two types of proteins combine to form a nanoscale morphology, which is mainly composed of ordered (crystalline) and disordered (amorphous) polymers, at the smallest level. The ordered domains can be attributed to oriented beta-sheet crystals with strong amide–amide hydrogen bonding, and the disordered domains to a wide range of structures with different degrees of hydrogen bonding.[8–9] Due to the heterogeneous complex, the structures of domains are kept within nanoscale dimensions. It allows energy throughout the material to transfer with high efficiency (i.e., rapid and comprehensive). Perhaps counter intuitively, silk has a relatively low stiffness (5–15 GPa) for a high-strength fiber. Quantitatively, spider silk has a maximum strength of ~1.7 Gpa. It is estimated to require domains of the order 2–4 nm in size, which can compare "grain dimensions" to the units of strength in super-hard metals and ceramics.

Unlike other important biological materials, silk results from continuous extrusion but not the cellular growth processes that produce collagen, chitin, or keratin. As polymer extrusion is an industrial process, while the controlled structure in industrial growth is yet not a reality, we can make silk as a template to design and produce a functional biomaterial. We have explained that the nanoscale interactions with structural simplicity but functional complexity make silk have desirable and highly tunable properties. This is a complex processing procedure. The protein molecules fold containing both intramolecular self-organization and intermolecular self-assembly.[10–11] Genetic control of the exact positioning of the amino acids guide these two processes. Meanwhile, the amino acids also make up the side chains of silk-polymer macromolecules. Factually, it is also a question that whether the true liquid crystals and semicrystalline complexes are formed by the block copolymer molecules of the individual silk proteins. But the synthesis of the primary structure of the protein is considered to generate a precursive morphology.[12]

The spider's silks are linked to the biological rule of the macromolecular chemistry of silk proteins. This link is realized via the physics storage and mechanical energy dissipation at a molecular level. Here, we only outline the same in order to illustrate the elegant design of natural proteins, which allows major changes in property. This property can be controlled by apparently structural changes. More recent work has also developed some models to show how water affects the properties of silk. These models give insights into the mechanical properties of other biological polymers that make a function in wet environments. This method suggests that we can give an idea to predict the full range of properties that are possible in most natural polymer materials.

The first problem to create a model to show the protein function is the need for a quantitative parameter to describe the structure of these proteins. We have known the peptide sequences of many silks and many different secondary structure types. However, the detailed crystallographic structures in the different silk types have not established a clear consensus. Now, we use "order" and "disorder" peptide segments fractions to quantify the morphology of specific silk, rather than define structure in terms of specific space groups. There exists difference between "order" and "disorder" segments. The "ordered" segments have two hydrogen bonds between amide groups. In contrast, the "disorder" segments have only one hydrogen bond with per amide–amide interaction because of misalignment of adjacent chain segments. The complex secondary structures play an important role in determining mechanical properties.

People always want to get the artificial spider silk through large-scale preparation of the same nature with natural spider silk. Dupont said that to obtain high-performance artificial spider silk, controlling the assembly and structure of the material from molecular level is needed. A practical method is the recombinant technology. After a few years study, they synthesized some genes and hoped to realize the expression out of spider silk protein. These genes had been implanted into the yeast and bacteria, which produced a replica of silk proteins. However, the protein folding and assembly process cannot be well controlled. So the mechanical properties of these proteins were not comparable with spider silk. This field of research is attracting more and more scientists to be involved.[2,11,13–17]

3.1.2 The Influence of Water on the Mechanical Properties of Spider Silk

Spider silks can provide a model as an ideal natural material for a wide range of biological elastomers, for it can function in states of hydration. This is because of the wet capture strands from dry structural threads of a web. Water has an important effect on the mechanical properties and elastomers of spider silk.[18–22] It is a key requirement to understand the interaction of biological elastomers, if we want to produce the biomimetic analogues. After all, the biological functionality tends to rely on wetting engineering.[12] Most typical bio-materials function inside the body or at the interface between body and environment. However, spider silk is different in working from the animal. Typical silks function either in the dry state or only partially hydrated state. As all the materials in the body are always bathed in water, they have evolved both functionality and molecular architecture for that fully hydrated stage. However, spider silk encounter a wide range of hydration in its natural state. It provides a rare opportunity to study the effect of water on all functions of spider silks.

Most spider dragline silk outperform other natural fibers and most man-made filaments. Their sensitivity to water is an important characteristic of different silks. Supercontraction defines that some silks shrink considerably upon exposure to water. Absorbed water makes significant shrinkage for an unrestrained dragline fiber and leads to reversible conversion of the material into a rubber. This process is a functional adaptation for the silk to play a role in the spider's web. Specific motifs in the silk proteins and entropy-driven recoiling of molecular chains can control and induce the supercontraction. Similarly, the mechanical properties can induce changes by man-made fibers thermal shrinkage, which is attributable to the entropy-driven disorientation of "unfrozen" molecular chains or the "broken" intermolecular hydrogen bonds. Liu et al. showed how nephila major-ampullate silk affect the interplay between shrinkage and mechanical characteristics in a spinning condition. This interaction reveals some design principles linking the silk's molecular orientation to the exceptional properties.[11]

While shrinking, silks show a large drop in their elastic modulus and an increase in their strain and this effect is lost when they are dried. Major silks are sensitive to water. It is an important scientific question to understand this selective response to water. It helps to understand silk properties overall and deploy spider's silk as implantable biomaterial used in commercialization.

The key step is to quantify the effect of water for modeling most structural proteins. Because water is not only in natural environment, but also principal plasticizing agent of biology. Vollrath et al.[3] discussed that water influences only on the disordered segments. And it is reasonable to assume that water molecule has hydrogen bond to disordered amide groups. The authors took keratin for example, for it is a biomaterial to function in the dry state and outside the body. Semicrystalline of keratin is composed of 0.3–0.4 fraction of ordered crystal fibrils in a disordered matrix. We predict the isotropic tensile modulus of dry keratin ranges from 5.5 to 6.5 GPa, unchanging at 2%. When increasing levels of relative humidity (RH) and hydrating the disordered matrix fraction, the hydrated segments have a loss. This shows that modulus reduces with hydrated fraction until all the disordered segments have been hydrated at 100% RH. Then it reaches the lower limit of between 0.5 and 1 GPa. However, unlike keratin, collagen has evolved to work in the wet state and inside the body. But it is also a semicrystalline material with fractions of order and disorder. Like keratin, the predicted dry modulus is about 6 GPa. And as measured, the hydrated modulus is again about 1 GPa. Collagen is very important, and much more widely spread. Biopolymer makes up over 80% of our body's weight. Accordingly, silks with collagen behavior or conducive to collagen infiltration are of great interest to bio-medical silk researchers. Elastin has an even more complex structure than collagen. It is an important biopolymer with a range of biological functions. One of these is to keep our blood-pressure in healthy range through its action in arteries. Elastin has a rubber-like modulus of about 1 MPa. Taking hydrated elastin to be fully disordered, the calculated bulk modulus is 3.5 GPa. Those illustrate that the humidity or water has an effect on the properties of silks.

The structure and properties of these important bio-functional polypeptides such as collagen and elastin have been studied extensively with many exciting developments in recent years. Nevertheless, at present their chemical composition and physical morphology still confuse the quantitative explanation and accurate prediction of mechanical properties. It would be very useful to guide our studies on spider silks. The structure–function relations can also give us a good idea to understand other natural biomaterials. The materials have evolved to interact with the environment. Due to the interaction with the body physiology, they are difficult to study quantitatively.

So, spider silk is an amazing material in nature. It shows many properties in mechanics, elasticity, and the influence of water. So we often compare spider silk with other biopolymers. Elastomeric spider silk has a modulus of the order 1 MPa, which has the mechanical properties of a rubber-like polymer. This web capture-silk is composed of a flagelliform core fiber and an aggregate coating. The coating provides hygroscopic components that keep the silk fully hydrated even under the quite dry ambient conditions. The deeply hydrated silk material can even be compared with mammalian elastin in mechanical properties.

The capture silk of Ecribellates can form a viscous droplet covered by aqueous solution. And the air humidity also has a great influence on the formation of sticky droplets. When the RH is 9%, the sticky droplets formation requires several hours. And when the RH is 45%, it only needs a few minutes to achieve. While the water content is more than 80% on sticky droplets, it can plasticize core threads to have elasticity. The sticky droplets use the viscosity and surface tension of the liquid to curl up the loose threads, which likes wheels to tension the core threads. The result is that spider web will not slack without impact. The excellent flexibility of capture silk is principally derived from the water film and droplets on the silk surface. The main components of the water film are serine, betaine, choline, acetyl ethanesulfonic acid, and γ-aminopropyl butanamide. They have a function to well protect and stabilize the protein, which is conducive to the formation and stabilization of the sticky droplets. And they may have the function of killing bacteria and affecting the nervous system of insects.

Water also has an impact on the dragline silk. After the dragline silk is wetted by water, it will shrink, at the same time, doubling the diameter than the original. The difference of mechanical properties before and after wetting and shrinkage is derived from the impact of water on the internal β-crystal. Before wetting, the major axis of the crystal is arranged in the direction of silk, having a very high tensile strength. After wetting, the crystal renders a random accumulation distribution inside of the silk, decreasing the tensile strength.[23,24]

Indeed, full understanding the natural silks can lead us to design the material similar with silks. Or linking the spider silks to other bio-fibers, perhaps we can realize the important biological function with excellent mechanical properties. Spider silk can provide a mechanical property that extends upward in modulus by mineralization in materials such as bone. Porter[12] has published a model for structure–property relation that is consistent with the model for silk, so the same underlying mechanisms operating in silk can be used as a general template for understanding a wide range of biomaterials.

In conclusion, spider silk has been around for almost 400 million years as an extraordinary natural material. Its history is even longer than dinosaurs. Many silk proteins can retain for a long time, which is related to the essential materials properties of spider silks and spider webs. Scientists have proposed the models to illustrate the molecular and the multimolecular structures of capture silk. Because capture silk is found only in orb webs and is thus difficult to isolate from dragline silk. It is hard to investigate. But we will introduce some new property about the wetting capture silks as follows.[4]

3.1.3 Directional Water Collection (and Mechanism)

Superhydrophobic surfaces defined for surfaces with a water contact angle (CA) above 150°, have found a wide range of application and attracted a great deal of attention. Due to small contact area with water, it limits both chemical reactions and bonding formation through water on a superhydrophobic surface. Accordingly, such as the adherence of snow or raindrops, oxidation, and friction drag, these phenomena are expected to be reduced or inhibited on such surface.[25] Superhydrophobic surfaces require both low surface energy and appropriate surface roughness. Many methods to realize these two requirements have been reported. Before introducing water collection on silks, we should introduce some knowledge of superhydrophobic surface at first.

The classical Young's equation has described a CA θ of a liquid droplet on a flat solid surface, as follows:

$$\cos\theta = \frac{\gamma_{SV} - \gamma_{SL}}{\gamma_{LV}} \tag{1}$$

where γ_{LV}, γ_{SV}, and γ_{SL} are the interfacial free energies per unit area of the liquid–gas, solid–gas, and solid–liquid interfaces, respectively. Young's angle is a result of thermodynamic equilibrium of the free energy at the solid–liquid–vapor interphase. Depending on the value of the CA, surface properties are determined as hydrophobic (CA > 90°) or hydrophilic (CA < 90°). In practice, two types of CA values are used: static and dynamic CAs. For a flat surface, static CA is close to Young's angle. Dynamic CAs are nonequilibrium CAs. Static CAs are obtained by sessile drop measurements, where a drop is deposited on the surface and the value is obtained by a goniometer. Dynamic CAs are measured during the growth (advancing CA, θ_a) and shrinkage (receding CA, θ_r) of a water droplet. The difference between θ_a and θ_r defined as CA hysteresis ($\Delta\theta$). The values of $\Delta\theta$ can be as low as 10° for a self-assembled monolayer on silicon, and many surfaces show much larger hysteresis due to chemical heterogeneity and roughness. It is known that surface hydrophobicity is enhanced by surface roughness, according to two different equations.

The superhydrophobic surfaces are usually covered with micro- or nanoscale structure. This composes the roughness. Water can either penetrate the asperities or suspend above the asperities. In another case, in this surface, much higher CAs are observed. These two situations are defined as the Wenzel state (penetration) and the Cassie–Baxter (CB) state (suspension) for two related models: the Wenzel and CB models. These models form the basic guidelines for the study of superhydrophobic surfaces.

Wenzel equation is a theoretical model to describe the CA θ' at a rough surface. Wenzel modified Young's equation as follows:

$$\cos\theta' = \frac{r(\gamma_{SV} - \gamma_{SL})}{\gamma_{LV}} = r\cos\theta \tag{2}$$

where r is a roughness factor, defined as the ratio of the actual area of a rough surface to the geometric area. This factor is always larger than unity. In Wenzel's equation, the surface free energy of the solid part of a rough surface is r times higher than that of a flat surface. And increasing of the solid–liquid contact area can enhance the hydrophobicity of a rough surface. The actual surface area divided by the projected surface area is defined as roughness factor, for a rough surface $r > 1$. Wenzel predicted for a hydrophobic surface $\theta' > \theta > 90°$ and for a hydrophilic surface $\theta' < \theta < 90°$. Depending on the nature of the corresponding flat surface, roughness enhances both hydrophobicity and hydrophilicity. In this model, both the CA and its hysteresis increase with the roughness factor increasing for a hydrophobic surface. At the same time, it has demonstrated that the CA increases steadily with the roughness factor. When it exceeds 1.7, the CA hysteresis starts to decrease, which is contrary to the Wenzel's prediction. The decrease in the CA hysteresis is result of switching from the Wenzel to the CB state. The increasing air fraction leads to the suspension of water droplet on the top of the asperities.

Cassie and Baxter[45] proposed an equation to describe the CA θ' at a heterogeneous surface with two different materials. When a unit surface has a fraction f_1 with a CA θ_1 and the other fraction f_2 with a CA θ_2, the following equation can express the CA on the surface:

$$\cos\theta' = f_1 \cos\theta_1 + f_2 \cos\theta_2 \tag{3}$$

They applied this equation to the CA θ' of a rough hydrophobic surface. Assuming the surface composed of solid and air, the surface is trapping air in the hollows of the rough surface. There is $f_1 + f_2 = 1$, when f_2 represents the fraction of trapped air, Eq. 3 can be modified as follows:

$$\cos\theta' = f \cos\theta + (1 - f) \cos 180° = f \cos\theta + f - 1 \tag{4}$$

where f is a fraction of the solid–liquid interface and (1 – f) is that of the solid–air interface. In Cassie's equation, the decrease of the solid–liquid contact area can emphasize the hydrophobicity of a rough surface.

In conclusion, theories can qualitatively predict the CA of a rough surface. However, it is not evident to choose which theory to be used and when. Therefore, it is very important to get some guidelines to predict the surfaces behavior. It is critical in designing superhydrophobic surfaces. Thus, many researchers have an interest in producing model superhydrophobic surfaces.

Johnson and Dettre[44] made a model to show the variation of the water CA of hydrophobic surfaces with various roughness extents. They assume idealized sinusoidal surfaces. When the Wenzel's mode is dominant on roughness regime, they illustrated that the CA and hysteresis on hydrophobic rough surfaces increase with the roughness factor increasing. They also showed that if the roughness factor exceeds a certain level, the CA increase continues but the hysteresis starts to decrease. Hysteresis decreasing occurs as a result of the switching the hydrophobicity mode from Wenzel's to Cassie's because of the increase of the air fraction at the interface between water and solid.[25] Superhydrophobic surfaces show a promising application in microfluidic systems. It is from their water-repellent and low-friction properties over recent years. Recently, experiments have realized the designed microstructures applied to direct the droplet motion and construct wettability gradients.[26]

Over the past decade, the rapid improvements of micro-/nanofabrication techniques have allowed for the development of microfluidic systems. Due to system miniaturization and integration, a series of bioassays and chemical analyses can now be carried out on a square-centimeter-large chip with low cost and high efficiency. In particular, researchers have strong interest in droplet-based microfluidics in academic and industrial fields in recent years. Discrete microdroplets of isolated and confined reagents compare to traditional continuous fluid flow systems. They have fast mixing velocity, dispersion-less transportation, and small convecting volumes. Microdroplets are the ideal chemical reactors and have promising applications in microfluidics field. However, in microscale systems, it is difficult to control the tiny droplet to transport in a precise way. Hence, the droplet motion is eager to develop and plays an important role in directing the microdroplets on microscale. Especially, considerable approaches of wettability gradient actuation or surface tension have been introduced to control the motion of droplets. Genzer et al.[46] have recently reviewed this field in detail. However, conventional methods involved in thermal, chemical, electrostatic, and optical principles, which may lead to a new problem: whether the compatibility of reagents can be kept with the change of biochemical or biophysical properties. Furthermore, as a result of the nature of the contact, the delivered materials might be adsorbed on the solid walls. This will lead to mass loss in the process of liquid contamination and transport or clogging of channels.

Inspired by the lotus effect, superhydrophobic surfaces have been extensively researched in recent years. Now we have recognized that surface roughness is important to create superhydrophobic surfaces or form composite states and fakir droplets that is suspended on the surface roughness. Due to the small CA hysteresis and very large CA, superhydrophobic surfaces have ideal droplet-sliding and liquid-shedding properties. Microfluidics and no-loss transport may get great potential applications from the unique wettability. Microstructures are designed using controllable or tunable superhydrophobic surfaces, which can direct the motion of droplets. Exploratory research about this has been carried out. These researches may greatly improve the development of microfluidic systems. For example, Sandre et al.[47] created a surface roughness as ratchet-like topographical structures and originally introduced liquid droplets directional. He reported the droplet motion across the boundary between flat and rough surfaces. Moreover, Petrie et al.[48] first introduced porous substrates coated with chemical wettability gradients, which can reduce friction at the liquid/substrate interface. The microdroplets can move directional along the substrates. Following this work, Shastry et al.[49] successively showed the manipulation of droplet motion. In addition, Yang et al.[50] reported droplet manipulation on wettability gradients that are actuated by the alignment of parallel patterned microtextures.

Seen from the background, many biological surfaces, including plant and animal kingdoms, owe unusual structural features of the micro- and nanometer scale. This special structure controls their interaction with water and lead to wettability.[1,27] These special structures determine the interaction with water, which exhibits wonderful wetting ability. For example, water striders,[1] lotus leaf,[28] desert beetles,[29] cicada wings,[30] fish scales,[31] butterfly wings,[32] etc. A very interesting example in our lives is the Namib Desert beetle, whose back use micrometer-sized patterns of hydrophobic and hydrophilic regions. The beetles use these patterns to capture water from the air to meet its subsistence need. Another example is the spider silk that can collect water from the air. Anyone who has seen a spider web in the early morning dew will notice that water droplets string along its fine threads (Fig. 3.5). This phenomenon attracts Zheng et al.[18] to research "how does that happen." They did an experiment with a human hair, unfortunately, water would not stick to it like that. So this phenomenon makes them to guess that spider silk has a strong ability to capture water from the air. The spider silk is composed of humidity-sensitive hydrophilic protein and gets an extensive attention due to its excellent mechanical properties. However, this interesting performance has rarely been studied, which is the ability to collect water from the humid air. So Zheng's team launched an almost five-year-long study with the initial curiosity.

Figure 3.5 Shiny water droplets on a spider's web.

They introduced a kind of *U. walckenaerius* capture silk. The reason why this spider silk has water collection ability is a unique fiber structure after wetting, with the "wet-rebuilt" fibers characterized by periodic spindle knots made of random nanofibrils and separated by joints made of aligned nanofibrils. In addition, the spindle knots and joints, respectively, are covered by a random arrangement but a certain degree of alignment of the nanofiber. Because of the different roughness of spindle knot and joint, the surface energy gradient is produced on these two positions. At the same time, due to the different curvature of two positions, there is a difference of Laplace pressure. Under the cooperation of two different driving forces, micrometer-sized water drops can achieve continuous condensation and directional collection of water drops around spindle knots. In addition, the stretch orientation of the nanofibrous structure makes it easier about the drops movement in the orientation direction. The movement of submillimeter-sized liquid drops has been reported by the surface energy gradients[33,34] or the difference in Laplace pressure. But for micrometer-sized drops, it has a larger hysteresis effect, which makes the movement more difficult. But, until now no force has been used to overcome the larger hysteresis effects to make the micrometer-sized drops move. However, the spider silk has made it come true to drive micrometer drops by tapping into both driving forces. The findings could have implications for the design of materials for the efficiency of chemical reactions and for water collection.

Spider silk is famous for its excellent mechanical properties, and the fact that it is composed of humidity-sensitive hydrophilic flagelliform proteins. In this chapter, we pay more attention about the ability to collect water from humid air. As shown above, cribellate spider can build a web using a cribellum—a comb-like device—to separate silk fiber from its spinnerets into many exceedingly fine fibers. Spider silk is composed of two main-axis fibers with some puffs and nanofibrils. The periodic puffs and joints surround two main-axis fibers, and the random nanofibrils are spaced on the puffs. The diameter of puffs and joints is measured by the environmental scanning electronic microscopy (SEM), and the value of them is 130.8 ± 11.1 μm and 41.6 ± 8.3 μm, respectively. The nanofibrils diameter ranges from 20 to 30 nm. Because the nanofibrils are hydrophilic protein that can enhance the wettability of spider silk, so the nanofibrils can help water drops condense.

We have discussed about the humidity sensitivity and rebuilt structure of spider silk. What is interesting is that the structure of spider silk will change when a dry spider silk is placed in a high humidity. Figure 3.6 shows the SEM images of wet-rebuilt spider silk. We can see that wetted spider silk has a more obvious spindle knots and joints structure. Figure 3.6a shows the apex angles 2β is about 19° and the periodicity length is 89.3 ± 13.5 μm. The diameters of spindle knots and joints of wetted spider silk also changes into 21.0 ± 2.7 μm and 5.9 ± 1.2 μm, respectively. The low-magnification and high-magnification images are added, respectively, to show the structure of spindle knots (Fig. 3.6b and c) and joints (Figs. 3.6d and e). Both spindle knots and joints are composed of randomly nanofibrils. But there are some differences between spindle knots and joints nanofibrils. The nanofibrils of spindle knots give a rough surface, but these of joints form an anisotropic aligned. Compared to the spindle knots surface, the surface of joints is smoother. This structure features form two water drops driving forces, surface energy gradient, and a difference in Laplace pressure. This is why the water drops can move directionally. We will give an obvious experiment phenomenon about the water coalescence and movement along the silk fiber as followed.

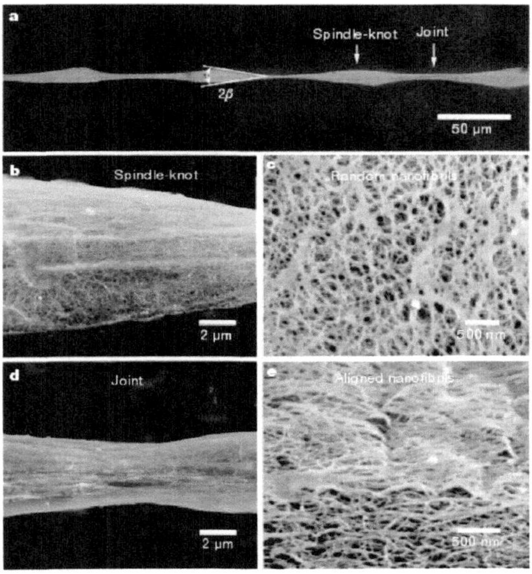

Figure 3.6 Structure of wet-rebuilt spider silk.

We put the dry spider silk in high humidity. The silk structure will change, at the same time, some water condense on the spindle knots and joints. The formed drops move toward the spindle knots. First of all, Fig. 3.6a shows the water drops (indicated by arrows) appear on the puffs. Then the puffs shrink and form spindle knots and joints with water drops growing (Figs. 3.7b–d). We have explained the influence of fog on the mechanical properties of spider silk in detail in above content. This research further illustrates that water wetting have changed the structure of dry spider silk, which leads to the mechanical properties change. But the re-built structure of wetting spider silk is the start of water collection. We divided the images into four regions to give a detail to explain the directional water collection on the wetting spider silk (Figs. 3.7e and f). In Fig. 3.7e, we see the small water drops condense on the spindle knots and joints. The arrows and water drops number show the water drops directional movement from the joints to spindle knots. At last, larger drops L, M, and N cover on the spindle knots. The three larger drops are coalesced, respectively, by 1–10 small water drops. Figures 3.7g–i show more details about water collection on an individual spindle knot. We can observe clearly that small water drops 1′, 2′, 3′, and 4′ condensed on the spindle knot and two joints at 0.156 s. With the drop 3′ growing, it moves toward to drop 4′ and condense with drop 4′ to form drop H. At 0.702 s, the growing drops 1′ and 2′ move directionally to the middle drop H. A larger drop H′ is formed by drop 1′ and drop 2′ condensing with drop H at 0.796 s. Then new directional water drops (1″ and 2″) start to cycle this process of water collection. The black arrows indicate that the directional water collection will continue with time.

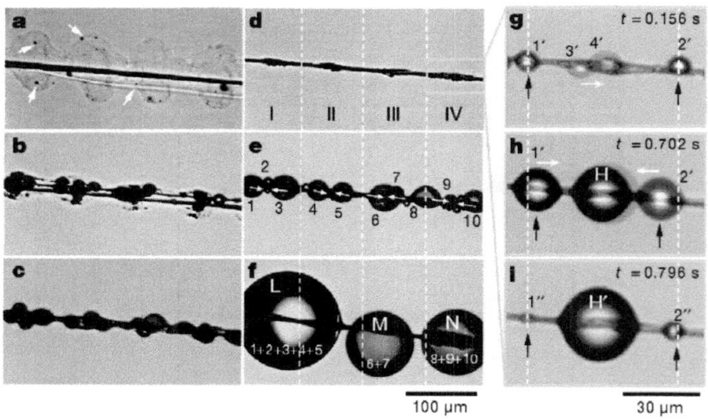

Figure 3.7 In situ optical microscopic observation of directional water collection on spider silk in mist.

In this process, the spindle knots play a role of condensing site in the initial stage and collecting site for the smaller drops movement directionally from the joints. And the joints serve only as condensing sites. The water condenses on the joints and grows into drops. Then they are transported from the joints to spindle knots. What it means is that this is a cycle process. When water drops complete condensation and collection on the spindle knots and joints, the second process will start immediately. This will promise continuously to use spider silk to collect water. These images show us the cooperation between joints and spindle knots acting as different sites, condensing sites and mainly as collecting sites, respectively, to realize the directional water collection. Why does spider silk have the ability to collect water? In fact, the silkworm silk and uniform nylon fiber cannot exhibit this property. Relate to the SEM images of wetted structure of cribellate spider *U. walckenaerius* silk we have illustrated in Fig. 3.6, maybe the answer can be got. We have shown the roughness difference of spindle knots and joints surface, which leads to surface energy gradient and a difference in Laplace pressure. Here, we will give more detail about the mechanism of spider silk collecting water. There are two reasons to affect the surface energy gradients, different in surface chemical composition and surface roughness. And surface energy gradient is one of the forces that drive water drops movement from low surface energy to higher one on the spider silk. We know that Wenzel's law is

$$\cos\theta_{\mathrm{W}} = \gamma\cos\theta \tag{5}$$

where γ is the surface roughness, θ_{W} and θ are, respectively, the contract angles on different surfaces. As spider silk is composed of hydrophilic protein, the chemical composition on the spindle knots and joints is the same. But in Fig. 3.6, we have shown that joint has a smaller axial–parallel roughness, which leads to a larger CA than spindle knots. Namely, the spindle knot is more hydrophilic than joint with a higher surface energy. So a force is given by the surface energy gradient from a different surface roughness, according to

$$F = \int_{L_j}^{L_k} \gamma = \cos_{\mathrm{A}} - \cos\theta_{\mathrm{R}})\mathrm{dl} \tag{6}$$

where γ is the surface tension of water, θ_{A} is the advancing angle and θ_{R} is the receding angle of water drop on spider silk, dl is the integrating variable from joint (L_j) to the spindle knot (L_k). The less hydrophilic part at joint has lower surface energy, and the more hydrophilic region at spindle knot has relative high surface energy.

The other driving force is a difference in Laplace pressure for directional water drops movement from joints to spindle knots. As shown in Fig. 3.5b, the spindle knot is a conical object. And a conical shape can give a different in Laplace pressure (ΔP) because of a curvature gradient. According to the formula:

$$\Delta P = -\int_{r_1}^{r_2} \frac{2P}{(r+R_0)^2} \sin\beta dz \quad (7)$$

where $\gamma$is the local radius, $R_0 = \left(\frac{3V}{4}\right)^{1/3}$ is the drop radius (V is the drop volume), β is the spindle knot half apex-angle, and z is the integrating variable along the diameter of the spindle knot. The joint has a local radius r_1 with high curvature site, which has a larger Laplace pressure. Spindle knot has a local radius r_2 with low curvature site. As r_1 is smaller than r_2, so the spindle knot has a relative smaller Laplace pressure. The result of different Laplace pressure will propel the drop movement from joint site to the spindle knot. So the surface energy gradient is from anisotropic surface structure and the difference in Laplace pressure is from the conical spindle knot geometry. So the cooperation of surface energy gradient and difference in Laplace pressure results in directional water drops movement from joints to spindle knots.

In fact, the significance of this finding is that the two driving forces can move a small drop with the diameter of micrometer scale. A submillimeter-sized liquid drop is driven by one of two forces mentioned above, which has been reported before. But a micrometer sized water drop has an increasingly important contact hysteresis effect. So the unique structure of wetted spider silk combines Laplace pressure differences and surface energy gradients to overcome the hysteresis effects of a micrometer-sized water drop. A comparative test also proved that the ability of directional water movement will disappear without the anisotropic structural features. Indeed, previous reports have shown some methods to drive liquid drop to move by one of two driving forces mentioned above.

Here, we introduce a research about only gradient driving water drops. The Marangoni effect is famous as an important surface phenomenon. It drives liquids movements by unbalanced surface tension forces. These types of flows have the valuable industrial applications, such as design and operation of integrated DNA analysis devices and microfluidics. Various conditions, such as composition and temperature on a liquid surface can lead to Marangoni motions. It also can be triggered by surface tension heterogeneity on a solid substrate. The flow speeds range from micrometers to millimeters per second on a solid surface, which is too slow to have practical applications.

What stops the drops moving? On a solid surface, the main obstacle is from the CA hysteresis, which pin the drop edge. In order to reduce hysteresis, some other energy must be supplied to the drop. The additional force can be resulted from the gradient of surface tension, which biases the drop movement. Daniel et al.[33] reported a type of surface tension to guide flow. In this model, drops move faster than the speeds of typical Marangoni flows as hundreds to thousands of times. These configurations give us a guide of how this effect can be harnessed and then to design heat exchangers. The studies can be used in rapid phase change, from steam condensing on a gradient surface.

A diffusion-controlled silanization prepared the gradient of chemical composition in this study. Figure 3.8 shows the detail of the experiment. A alkyltrichlorosilane small drop of ~2 μL is held about 2 mm above the center of silicon surface. The silane evaporates from the drop and diffuses radially when it reacts with the silicon (Si/SiO_2) surface again. The central part of the silicon surface closed to the drop and became hydrophobic. The CA is more than 100°. However, the peripheral zone is wettable by water and the CA is near to zero. In the air, small water drops of ~1–2 mm move on such a surface radially toward the peripheral zone with speeds of 0.2–0.3 cm/s. However, when saturated steam (100°C) condenses on the surface, smaller drops of ~0.1–0.3 mm obtain speeds.

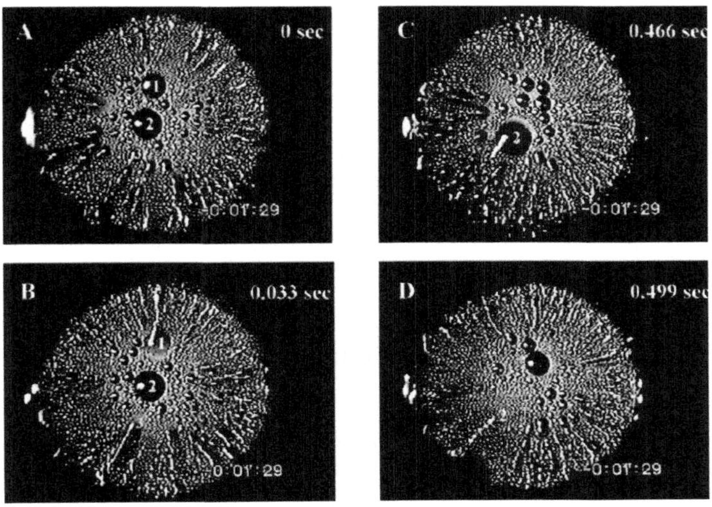

Figure 3.8 Video prints showing fast movements of water drops (indicated by the plume- and streak-like appearances) resulting from the condensation of steam on a silicon wafer possessing a radial gradient (1 cm diameter) of surface energy.

Such high speeds of drops attract researchers' attention. Hydrodynamically lubricating water film will make a loss of interfacial resistance, which can speed up drop motion. Some other plausible scenarios exist. The wetting hysteresis maybe bypassed under rapid condensation, or additional energy is supplied to surmount hysteresis for a drop. When a liquid drop is placed on a surface of energy gradient, Laplace pressure attempts to equilibrate quickly. So, Fig. 3.9 illustrates the mechanism. At edges A and B, a drop with the shape of a semispherical cap creates nearly equal CAs (θ_d). But, as θ_d is greater than the advancing CA (θ_{aB}) at B and less than the receding CA (θ_{rA}) at A, the unbalanced wetting force plays a role in both A and B. These forces propel the whole drop to move toward the region of higher wettability. The speeds of drops motion depends on hysteresis and drop size. The speed ranges of a few millimeters per second. Due to the presence of fast condensation, the direct condensation of steam and coalescence with other drops can aid to the drop movement.

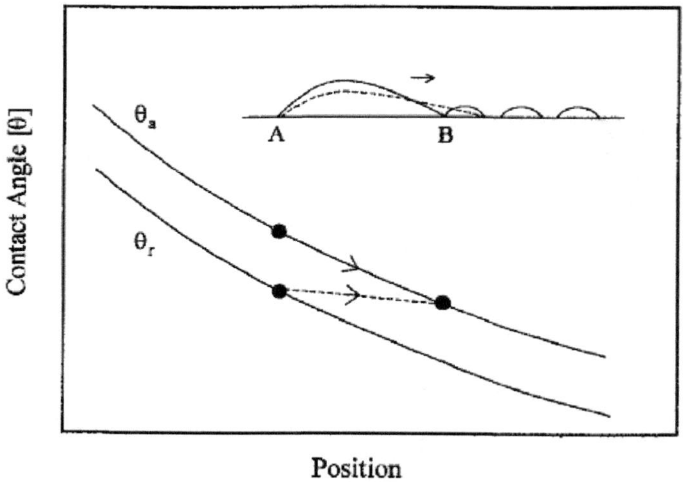

Figure 3.9 Schematics of a 1D wettability gradient of a surface. θ_a and θ_r represent the advancing and receding CAs, respectively. In the absence of hysteresis, the driving force for the drop motion is provided by the difference of the equilibrium CAs at points at B and A. Hysteresis reduces the driving force (dotted arrow). The upper right inset shows that additional driving force can be gained from coalescence with other droplets, which nucleate and grow ahead of the main drop.[41]

A different study is about Laplace pressure difference to drive water drops. Different ways have been described for asymmetric droplets movement. Bouasse found that a slug in a conical capillary tube could move toward the small radius region (Bouasse 1924[51]). Weislogel showed a drop motion because of the interface between two substrates with different chemical nature (Weislogel 1997[52]). Bain et al.[53] (1994) and Domingues dos Santos and Ondarc uhu[54] (1995) invented a dynamic version of drop movement, in which the motion itself keeps alive contrast in wettability between the front and the rear of the drop. Bico and Quéré[55] (2002) researched the motion due to an asymmetric slug in a tube. The contrast between the front and the rear of a drop can increase from a temperature gradient. Brochard[56] (1989), Smith[57] (1995) and Mazouchi and Homsy[58] (2000) described very completely about the drop motion with a temperature gradient. Carroll[59] (1989) focused on a drop placed on a mammalian hair fiber. He noticed that in the parallel direction to fiber axis drop had a tendency to move. This phenomenon is similar with spider silks. The difference of CA of liquid with the heterogeneous surface led to the transient movement.

Generally, capillary force is related to surface tension and characteristic length. If a drop is placed on a gradient of surface tension or characteristic length, it can be moved, which is described in detail above. Here, we show an example about movement of a drop on a conical fiber with radius gradient. In this case, the drop spontaneously moves toward lower curvature region. The drop velocity does not depend only on the conicity of the fiber but also on its local radius. The dynamics of droplets self-propelling on a horizontal conical fiber is not continuous. The speed of the drop decreases as the liquid moves. The position of fiber is also increasing with the drop volume. These features reveal to identify the viscous force that opposes the movement. This device allows us to control the driving force.[35]

Due to the combined effect of surface energy gradients and Laplace pressure difference, the wet-rebuilt spider silk can overcome the hysteresis effects so as to drive a micrometer-sized water drop directional movement from joints to spindle knots. As the spindle knots and joints have the same chemical composition, the different surface geometries affect the condensation and transfer of water drops. Drops move along the parallel direction of an aligned surface topography rather than a randomly rough topography surface. The reason is because of another influence factor, vapor–liquid–solid three-phase contact line (TCL). The TCL of parallel direction of the surface is continuous with aligned topography, and that of randomly rough topography is discontinuous. A continuous TCL will allow the drops spreading and moving, but a discontinuous TCL will give rise more significant hysteresis effects. We have introduced the composition of joints and spindle knots in wetted rebuilt spider silk, aligned nanofibrils, and random nanofibrils, respectively. So the TCL is continuous on the joint regions, and is discontinuous on the spindle knot regions. The water drops move along with joints with less hysteresis to spindle knots with more hysteresis effects. This is the third force to drive the water movement from joints to spindle knots.

3.1.4 Stronger Capillary Wet Adhesion (and Mechanism)

In order to minimize the interfacial free energy, a liquid drop on a chemically heterogeneous surface can move toward the higher wettability region. The wettability gradient can provide a driving force, which is compensated by the viscous drag force. Then the drop attains a steady velocity. The motive force increases with the area. And the drag force increases with the perimeter of contact. The steady-state velocity of the drop (defined in terms of capillary number, C_a) increases linearly with its base radius (R) as

$$C_a = \alpha R\left[\frac{\mathrm{d(cos)}}{\mathrm{dx}}\right] \tag{8}$$

where $C_a = V\eta/\gamma$; V, η, and γ are the velocity, viscosity, and drop surface tension, respectively, and θ is the position dependent CA. The coefficient R is a constant. It provided the drag forces at the contact line scale with the bulk viscosity. However, according to Eq. (8), all size liquid drops should spontaneously transfer toward the higher wettability region. Wetting hysteresis shows as a yield force at the TCL. Considering the CAs of a liquid on a real surface bounded by two extreme (advancing and receding) values, the situation can be understood. For a drop to move, its advancing edge angle must be smaller than its receding edge angle. This condition is assured with hysteresis-free gradient surfaces but not necessarily with nonideal surfaces.[34]

In recent years, larger hanging drops on a fiber have been paid more attention. It can be used in water-acquiring engineering or filtering projects. In above content, we have shown that cribellate spider capture silks can hang amazingly large pearly water drops after wetting. It displays a strong water capture ability on wetted cribellate spider's capture silk. However, this ability would not exist if there is special wet adhesion on a surface. We investigated the capillary adhesion during hanging larger pearly water drops. There is a special structure of spindle knots and joints on the wetted spider silk, which have different roughness and curvature. A novel model is introduced about capillary adhesion force to value the larger pearly hanging-drops. As-designed artificial fibers can further demonstrate the strong ability of water capturing. This investigation opens a wide insight into the wet adhesive property of spider silk. It is helpful to design artificial fibers to be applied into water collecting webs and tents. It also can be extended in filtering projects, such as the dust pollution from chemical plants and noxious emission of aerosol.[36]

The cribellate spider's capture silk has a series of "puffs," which are composed of a lot of nanofibrils. The numerous nanofibrils surround main-axis fibers. It will form spindle knots and joints structure after wetting, and they are composed of random nanofibrils and array ones, respectively. The spindle knots and joints contribute to drive the tiny water drops and realize water directional collection by the cooperation of Laplace pressure difference and surface energy gradient. Interestingly, we can see some cribellate spider's orb web hangs many larger water drops in the foggy morning. Experiments have observed an alluring water-capturing scene. A high-resolution observation by optical microscopy shows that the larger drops are stably hanged on spindle knots of silks. We will give a detail for this phenomenon in last section. Capillary adhesion of wetted spider silk plays an important role in hanging drops.

The special adhesion depends on the microstructures of spindle knots and joints, which is the base of the property existing. We have illustrated the "puff" of the viscid spiral of ecribellate spider silk and cribellate spider silk used to prey. But this property is different from that case. Zheng explains the capillary adhesion that exists due to the spindle knots of wetted capture silk. A novel model about the capillary adhesion has been established to show the reasons of larger pearly water hanging-drops. Based on the roughness and curvature, Zheng's team fabricates the artificial spider silks to demonstrate the wet adhesive ability using different methods, which we will introduce in detail in next section. This investigation on the adhesion ability of fibers is important in the design of novel materials. The novel material can be applied in filtering projects, water collecting, dust pollution from chemical plants, the emission of aerosol, and so on.

In order to explain the capillary adhesion, we further observe the geometry feature of the silk. We denote the apex-angle of spindle knots as 2β. The maximum diameter ($2b_r$) is on the center parts of the spindle knots, and the minimum diameter ($2b_a$) exists on the spacing parts between spindle knots. The rough feature of spindle knots result in puff-shrunk when hydrophilic nanofibrils absorb water in puff. The wetted spider silk results in the roughness (r), which is estimated by the ratio of $A_{true}/A_{apparent}$. Considering the roughness of spindle knots and joints (namely the spacing of spindle knots), A_{true} (true area) is far more than $A_{apparent}$ (apparent area). According to the Wenzel's law mentioned above, roughness plays an important role in increasing the wettability of a surface. Accordingly, it also can enhance the solid–liquid interface.

Carroll et al.[59] proposed a comprehensive framework with a drop on a thin fiber. They described how the capillary force (f_c) balances the gravity (G), when a water drop is hanging on the fibers. Quéré and Lorenceau[35] indicated that the increase of fiber radius leads to the hanging-drop enlarging its radius. We define that γ and ρ are the surface tension and liquid density, respectively. b_0 is the fiber radius. When $b_0 < \kappa^{-1}$, the capillary length $\kappa^{-1} = (/g)^{1/2}$ (κ^{-1} is 2.7 mm for pure water). So, the volume of drop is growing, and it will slide off or spread over randomly from a smooth fiber without any micro- or nanostructure. As a result, the drop cannot grow any more. So a larger pearly water drop cannot stably hang on a uniform fiber. The capture silk of cribellate spider can pin rich larger pearly water drops. Figure 3.5 have shown the larger drops are stably hanging on cribellate spider's capture silks. Figure 3.10 illustrates the wetted cribellate spider silk in detail. The rough curve features exist on the spindle knot (Fig. 3.10a). Due to the curvature and roughness, a larger pearly water drop can stably hang on the spider silk. The scheme (top) and optical images (bottom) of a hanging drop on horizontal silks are shown in Fig. 3.10b Frame 1–2. The maximum radius of hanging water drop on the horizontal silk is of ~0.95 mm. And, Fig. 3.10b Frame 3–4 shows the scheme (top) and optical images (bottom) of the hanging-drop on vertical silks. A maximum hanging drop of 0.54 mm in radius is obtained on the vertical silk (in Frame 4 of Fig. 3.10b). On a horizontal silk, the ratio of the maximum hanging drop volume to the spider silk region is estimated up to 52,000. Apparently, the capillary adhesion results in these larger hanging-drops. And, the roughness and curvature of the spindle knot on wetted spider silk can enhance the capillary adhesion.

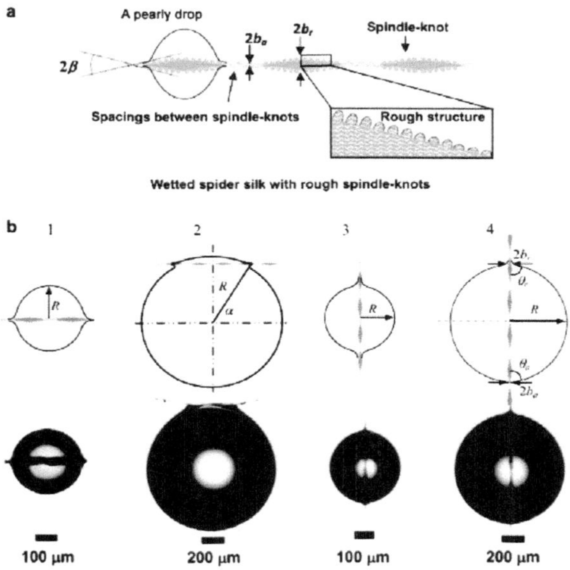

Figure 3.10 Illustration of hanging drops on wetted spider silk.

For a hanging drop on the horizontal spider silk, this situation is very similar to a drop hanging on a cylindrical nozzle whose radius is a. The drop gravity is

$$G = 4\pi \rho g R^3/3 \tag{9}$$

which can be balanced with surface tension $2a$. Lorenceau et al.[35] assumed a model about large drops hanging on uniform fibers. In this model, two forces are equivalent. One is from the horizontal fiber and the other is generated by two similar fibers connected to the drop. The capillary force can be calculated by ($4b_0\cos$ /2– α), where α is the off-axis angle, b_0 is the radius of fiber. But the wetted spider silk is more complex because of the rough curved spindle knots. The roughness factor should be considered to enhance the surface adhesion according to Wenzel's law. The roughness should be introduced to describe the capillary force for a hanging drop as follows:

$$f_{c,H} = r \times 4\pi b\ \gamma \sin\alpha \tag{10}$$

In this formula, $f_{c,H}$ is the capillary force of horizontal silk; r is the roughness, and b is the effective radius of silk. When a middle-sized larger drop covers on one spindle knot, the α is a basically linear function of $\theta_b - \theta_u$, i.e.,

$$\alpha = 1.74665\ (\theta_b - \theta_u) + 0.61833 \tag{11}$$

where θ_b and θ_u are the below CA and upper CA, respectively. We know that the capillary force can be balanced by gravity of hanging drop. There is a relationship between sin and $R^3/3rb\kappa^{-2}$. However, as the water drops grow and coalesce, the off-axis angle α will be nonlinearly changed. Especially when an extreme larger hanging-drop is covering two or more spindle knots, the geometry of the spindle knot will limit the off-axis angle α. This is because the water drop is hanging stably on the spindle knots. Figure 3.11a shows the all kinds of size drops are pinning on horizontal silks, including a maximum pearly hanging-drop. The sin will be a nonlinear function of dimensionless quantity $R^3/b\kappa^{-2}$. Fitting of the testing date can obtain the relationship curve, as seen in Fig. 3.10b. There is a relation of

$$\sin\alpha = 0.10098 + 0.36771x - 0.03967x^3 \tag{12}$$

where $x = R^3/b\kappa^{-2}$. The scheme of a water drop hanging on spindle knots is inset in Fig. 3.11b. This research has indicated that the relationship of capillary adhesion and off-axis angle α is not linear for all kinds of size drops including maximum pearly water drops.

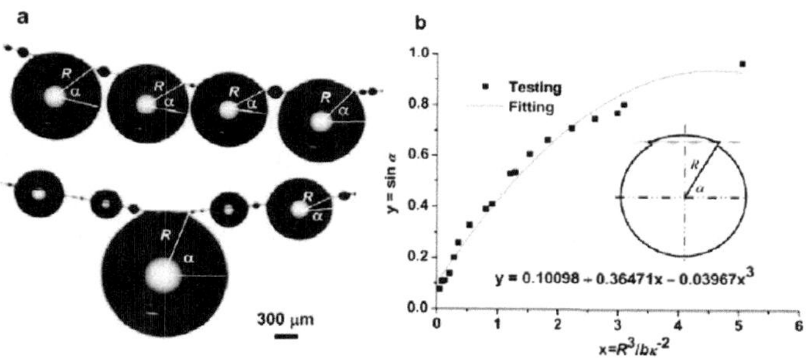

Figure 3.11 Hanging-drops of horizontal spider silk.

The Eq. (8) provides a basic formula to estimate the radius of a hanging-drop on a horizontal silk: R′ ~ $3rb\kappa^{-2}\sin^{1/3}$. Because of the curvature geometry of the spindle knot existing, we should consider the effect of the Laplace pressure difference. An item should be added to Eq. (8), which can improve the value of the maximum radius. The differential coefficient is considered, that is,

$$\Delta R = \Delta b \cdot r\kappa^{-2}\sin\alpha / R^2 \tag{13}$$

where $\Delta b = b_r - b_a$. ΔR can be estimated from the changed equation, so does the maximum radius of a drop hanging on horizontal silk with

$$R'_{H,max} = R'_H + \Delta R \tag{14}$$

Compared to the experimental result of optical image of Fig. 3.9b, the value of maximum radius is close. For a larger pearly drop hanging on a rough and curve fiber, the capillary adhesion can be described as

$$f_{c,H} = r \cdot 4\pi b\, \gamma \sin\alpha + \Delta_{Laplace} \tag{15}$$

where

$$\Delta_{Laplace} = [(1/3r) - B]\, x + Cx^3 - A \tag{16}$$

where $B = 0.36771$, $C = 0.03967$, $A = 0.10098$, as shown in Fig. 3.11b. This result indicates that the cribellate spider silk creates a strong wet adhesion to larger pearly water drops due to the structure of spindle knots with roughness and curvature on the silk. The capillary force results in a strong adhesion by spindle knots. And it appears the maximum volume ratio (~52,000) of the hanging-drop to the wet region of the spider silks.

Figure 3.12a shows the water drops hanging on the vertical silk. Compared to the drops hanging on horizontal silk shown in Fig. 3.12a, the size of drops are relatively smaller. The capillary adhesion is not related to the off-axis angle (α) of the hanging-drop. When the drop is growing, it slightly moves down along the silk-axis. It will stop until the top radius (b_r) of the drop is more than the bottom radius (b_a). When the top semi-angle (θ_r) of the drop is smaller than the bottom semi-angle (θ_a), the capillary adhesion force between the wetted spider silk and the drop is appeared. For the largest drop on the vertical silk, θ_r gets to the minimal value and b_r gets to the maximal value. According to the report complied with gravitational distortion of barrel-shaped drops on vertical fibers, the capillary force ($f_{c,V}$) for a drop hanging on the vertical silk can be described as:

$$f_{c,H} = 2\pi r\gamma(b_r \cos\theta_r - b_a \cos\theta_a) \quad (17)$$

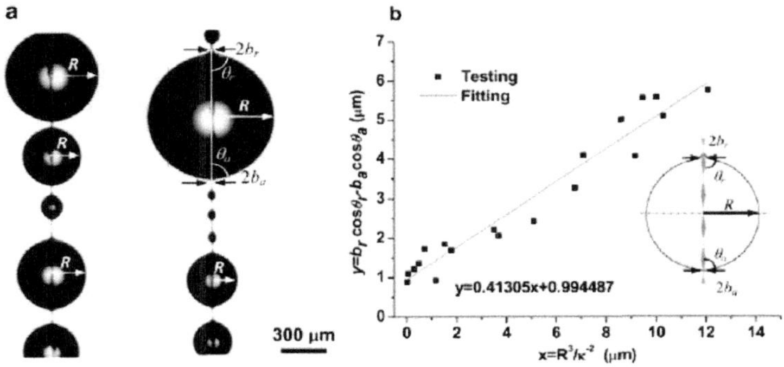

Figure 3.12 Hanging-drops of vertical spider silk.

where r is the roughness; θ_r and θ_a are, respectively, the top and bottom CAs of a water drop hanging on vertical silk, matching with top radii b_r and bottom radii b_a of spider silk, respectively. It shows that capillary adhesion is related to roughness, top/bottom semiangle, contact radii on the vertical silk, when a large drop is hanging on the spindle knot. The capillary force makes the drop hang on the spider silk stably. The capillary force is related to the difference of the top surface energy ($2b_r \cos \theta_r$) and bottom surface energy ($2b_a \cos \theta_a$) which can oppose to the gravity (G). The relationship of curve of the effective capillary force ($b_r\cos\theta_r - b_a\cos \theta_a$) and size ($R^3/\kappa^{-2}$) is shown in Fig. 3.12b. A linear relationship exists from the testing data. Geometric parameters of a drop hanging on vertical silk are inset in Fig. 3.12b. According to Eq. (17), by gravity balance, we can describe the maximum radius of a drop hanging on a vertical fiber as follows: $R'_{V,max} \sim [3/2.r\kappa^{-2}(b_r\cos\theta_r - b_a\cos\theta_a)]^{1/3}$, the $R'_{V,max}$ can be estimated. This result is near to the experimental value of the optical image as shown in Fig. 3.10b. The Eq. (17) indicates the capillary adhesion of the vertical silk to the water drop is also influenced by the cooperation of geometry and roughness of spindle knots of wetted spider silks. And, it also has a relationship with the top/bottom CAs and the effective contact radii of silk.

The novel models of capillary adhesion values the ability of water collection on horizontal and vertical wetted spider silk, respectively. Particularly, roughness and curvature can be proposed to value the capillary adhesion of spider silk. It opens a wide insight into a geometric-structured fiber. This study is helpful to design interface materials that can be applied in filtering aerosols; water collecting; detection and biosensors of nanobacteria, microreactors, and sample assays; and transportation of microdroplets.

3.2 Artificial Silk and Functions

There are many methods to realize the drop movement. In drop movement design, there are two key factors to consider, one is the driving force, and the other is the droplet-transporting capability. The surface gradient wettability has been applied to control drop movement. Chemical deposition with a surface energy gradient has also been used to create a transport path, which was reported by Chaudhury and Whitesides[60] in 1992. They further showed that this method could make a drop move uphill spontaneously. A drop runs from a hydrophobic region toward a more hydrophilic state on an inclined substrate. Grunze[61] fabricated a hydrophobic and hydrophilic pattern on complanate gold electrodes by self-assembled method. Daniel et al.[34] studied a novel cooling device by a radial distribution of the surface energy gradient. When the saturated vapor gas condensed, the cooling drops moved spontaneously and collapsed on the specific surfaces. Darhuber et al.[62] proposed a microfluidic device for liquid drops and continuous streams on a solid surface by integrated microheater arrays. This device could combine with liquophilic and liquophobic chemically patterned surface to serve as a logistic platform for the parallel and automated routing, mixing, and reacting of multiple liquid samples. Yasuda et al.[63] designed and refined a thin film with hydrophilic and hydrophobic surface by chemical deposition. Because there was a microchannel in the hydrophilic surface, the drops run gradually toward the other side of the region that was increasingly hydrophilic. Ichimura et al.[64] employed a light-driven method. A surface free energy gradient was proposed on a photoisomerizable monolayer. The liquid drop moved with UV light. These researches demonstrated that liquid drops can be delivered along a specified path by fabricating a patterned surface. There is a driving force to pull the drops motion.[37]

These discoveries reveal that various mechanisms to drive a drop movement are based on chemical gradient, thermal, and electrical driving or photo-responsive principles. However, there are some side effects when moving the drops in bioapplications, for example, temperature rise or chemical compatibility. Yang et al. generated a surface wettability gradients through microstructure distributions. This method based on Laplace pressure of a drop across distinct hydrophobic surfaces. This mechanism provides a concept that design of microfluidic systems can be realized by no undesirable side effect and no consumption of external power. Inspired by this novel idea, Yang et al. also created a liquid droplet separator and a passive microvalve. These creativities increased reliability for a microfluidic system. In these inventions, the surface was fabricated by a micromethod. And the drop contact behavior of this composite surface is similar to the lotus effect.

We have given more details about spider silks and the ability of water collection. Spider can be named as the biology's best "manufacturing engineers." It can fabricate lightweight, continuous orb-webs for prey capture. With a wetted environment at temperature and pressure, various types of spider silks can be applied for different purposes. These spider silks can stand wind, sunlight, and rain with strong mechanical properties. Spider silks possess micro- and nanoscale structures, such as the nanofibrils and nanocrystals. The nano- and microscale of spider silks plays an important role in achieving the excellent mechanical, chemical, and physical properties of spider silks. Because of the amazing properties of spider silk, many reports have been published to introduce the mechanism of spider silks. The other property of spider silk is exceptional elasticity. Capture silk is five times as elastic as dragline silk. Orb-weaving spiders depend on sticky capture silks to prey and absorb remarkable amounts of energy. So, spider silk is a multifunctional material with high strength, elasticity, and stickiness. Further understanding on how the spider silk is realizing the water motion. Large water drops hanging stably on the spider silks is an amazing phenomenon in the foggy morning. Water collection ability has been discovered as another fascinating property of the spider capture silk. The special hierarchical fiber structures are result of the water collection ability. The wetted fibers possess periodic spindle knots and joints made of random nanofibrils and aligned nanofibrils, respectively. The cooperation of surface energy gradient and Laplace pressure drives the water drops to move from the joint parts toward the spindle knots. Inspired by this study, some methods are introduced to design and fabricate a kind of artificial fibers with the structural features of wet-rebuilt spider silk. And, the artificial fibers can exhibit the water directional collection ability. This section will help in the simulating method of multifunctional materials for water collection. Overall, the spider silk presents multifunctionalities and it seems that the spider silk has yet many properties to be unraveled to the scientific world.

3.2.1 Dip-Coating Methods

The problem of water shortage can be solved by collecting water from fog air. This can be applied in dry and fog-laden areas. There are some creatures in nature with an amazing ability to collect water. Desert beetles and cribellate spiders are two examples. This ability is based on the special micro- and nanostructures on their surfaces. The desert beetle's back is composed of hydrophobic and hydrophilic regions. Due to gravity, water drops condense on the hydrophilic region and run away from the back. Some research has reported to artificially fabricating beetle's back. They mimic the beetle's back by fabricating a bio-inspired two-dimensional substrate with patterned wettability. The spider silks collect water by following a different mechanism. Spider silks not only own outstanding mechanical properties, but also have an amazing ability of collecting water from fog. It is well known that spider web often hangs numerous water drops in the foggy morning. Research has shown that the captured silk of cribellate spiders have a unique wet-rebuilt structure with spindle knots and joints. The cooperation between surface energy gradient and curvature gradient drives a tiny water drop (no more than 100 μm in size) to move directionally from the joints parts to the spindle knots. In our previous discussion, we successfully reported these structural features and the driving force about spider silks and water collection ability. These offered new insights in designing water collection materials.[38]

Inspired by the spider silk, we introduce three methods to fabricate a series of artificial spider silks with spindle knots and joints. The surface nanostructures or chemical compositions are subtly designed. Some investigations show that tiny water drops can be driven toward or away from the spindle knots. Controllable water direction can be optimized by the cooperation of roughness gradients, curvature, and chemical on the fiber surfaces. This section will help to find a way for designing smart materials and devices, which is applied in driving water drops in a controllable manner.

In this section, firstly, we will show the structures of spindle knots and joints fabricated under an optimized condition. A nylon fiber is immersed into polymer solution and drawn out horizontally. Due to the Rayleigh instability of the polymer solution, a string of polymer drops became spindle knots when being dried and formed spindle knots and joints structure on the fiber. Optimizing condition, these methods can control the size of spindle knots affecting the water collection ability. Secondly, because of the geometrically engineered thin fibers with much stronger water capture ability than previously materials, it results from an unusual TCL along the fibers. The TCL results from silk-inspired spindle knots structures on their surface. The hump structures combine the "slope" (spindle knots) and "curvature" (joints) to improve the stability of TCL. The humps provide sufficient capillary adhesion to hang the water drops. We will introduce the TCL length as well as the pinning effects. The characteristic geometry of the humps and the adhesion of fibers can improve the ability of water collection. These findings give a wide insight into solid–liquid adhesion of curve-structured fibers, as well as designing novel materials for promising applications in water collection microdevices or engineering.[39] Thirdly, the effect of geometry on water collection ability and the collection behavior is investigated by fabricating a series of the bio-inspired fibers with different sizes. Related to different of geometry on bio-inspired fiber, tiny water drops movement and the maximal volume of a hanging drop are discussed. The researches demonstrate that geometry of bio-inspired fiber leads to a much stronger water collection ability than that of uniform fiber. The geometry structures also are attributed to improvement of fog-collection efficiency by controlling the movement of tiny water drops to form large water drops at a geometric location. A new cycle of collection can improve the efficiency at the original place. This study is important for designing materials to develop the application on a large scale, such as fog water collection tents and web filtering projects of noxious emission of aerosol and dust pollution from chemical plants.[40]

3.2.1.1 Method

Here, we introduce a method named dip-coating. In this method, a uniform nylon fiber is firstly immersed into a polymer solution, such as poly(methyl methacrylate) (PMMA) or poly(vinylidene fluoride) (PVDF) dissolved in *N,N*-dimethylformamide (DMF), or other organic solution can be tried. We take the PVDF solution as an example. After the nylon fiber is drawn out horizontally using a dip-coater machine, a cylindrical PVDF film forms on the uniform fiber surface. Then the film spontaneously breaks up into polymer droplets owing to the Rayleigh instability. The bio-inspired artificial spider silks (BASs) are fabricated with periodic PVDF spindle knots on the nylon fibers after the evaporation of solvent.[39] Figure 3.13 shows that utilizing the Rayleigh instability method can fabricate the bio-inspired fibers with the interesting hump geometry, which are similar to spider silk. As explained in Fig. 3.13, a nylon fiber ($D \sim 18$ μm) is first immersed into PVDF/DMF solution of appropriate concentration (Fig. 3.13a). Then the nylon fiber is drawn out of the polymer solution at an appropriate rate (Fig. 3.13b). There is a solution film on the fiber surface. Figure 3.13c illustrates that the coated cylindrical liquid film instantly break up into periodic droplets along the fiber because of the Rayleigh instability. The geometries of the spindle knots can be controlled by solution concentration and the draw-out rate. The relationship of concentration and height of hump and that of draw-out rate and hump–hump distance will be given more details in the next content. Figure 3.13d shows the image of a typical bio-inspired fiber with a spindle knot height (H) of ~ 47 μm and hump-hump distance (λ) of ~ 400 μm.

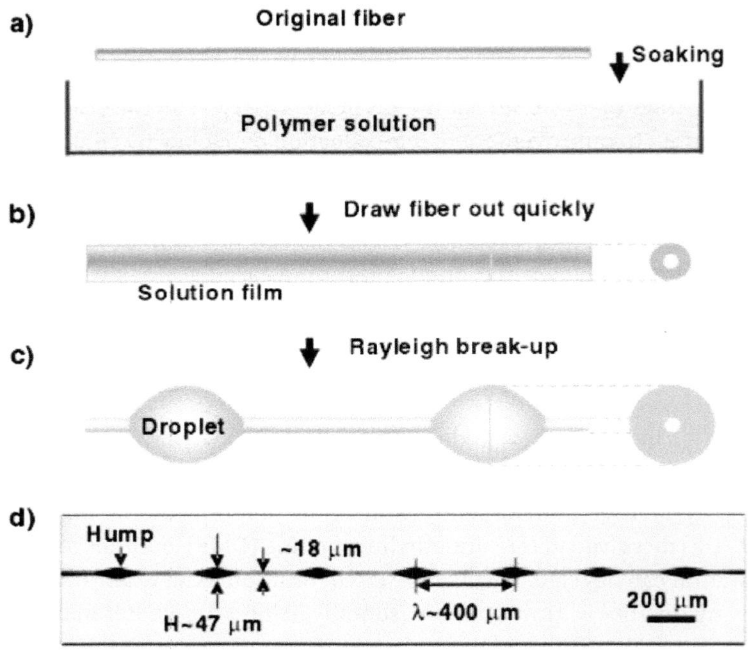

Figure 3.13 Scheme of fabricating bio-inspired fiber with hump structure.

An example is shown of the effect of solution concentrations and drawing out velocities on the size and periodicity of spindle knots. A series of BASs are fabricated under different solution concentrations and drawing out velocities. Figure 3.14 shows the optical images of them. Respectively, using 3%, 5%, 7%, 9%, and 11% (weight percentage of PMMA) of the concentration of the polymer solution and 10, 50, 100, 150, and 200 mm/min of drawing-out velocity of the dip-coater machine, the BASs can be divided into three different types based on the shape and size of spindle knots. In Fig. 3.14a, the knots have asymmetric shape, while in those in Fig. 3.14c, there are no obvious knots. Only in Fig. 3.14b, there are some periodic spindle knots. We can understand the above observations by theories developed for fiber coatings. The coating film thickness can be influenced by the capillary number (C_a) in a typical fiber coating method. This can be calculated as follows:

$$C_a = \eta V/\gamma \tag{18}$$

where η is the solution viscosity, V is the fiber drawing velocity, and γ is the solution surface tension. And film thickness is another factor influencing the formation of spindle knots. In this experiment, two steps are important for the formation of excellent periodic spindle knots. One is the enough PMMA solution on the nylon fiber when it is drawn out of the solution. The other is the enough time for the film to break up. The droplet by breaking up from the solution film must have enough time to form before it becomes dry and loses its flow-ability. The authors have measured the polymer solution viscosities that change from 5.42 mPa·s to 676.84 mPa·s and surface tensions changed from 28.14 mN/m to 36.16 mN/m with different PMMA concentrations. So the C_a values can be correlated with the fabrication conditions. For a low concentration (or viscosity), such as 5% and 7%, to form a thick film need a high drawing out velocity to make C_a large enough. But if the solution concentration is too low, such as 3%, no enough polymer solution is on the fiber. Even though the velocity is as high as 200 mm/min, the low viscosity (η = 5.34 mPa·s) results in a small C_a value. While, if the solution concentration is too viscous, such as 11%, before the film breaking up into droplets, the film has dried. Furthermore, the asymmetric spindle knots are formed due to the drag force for solution film from the highly viscous reservoir. These investigations show that we can control to fabricate periodic spindle knots on BASs by optimizing the conditions including solution surface tension, solution viscosity (or concentration), and drawing-out velocity. At a given velocity, a larger C_a results in more PMMA being drawn out from the thicker solution. So the size of the spindle knot increases with the solution concentration. However, there is a different relationship of size of spindle knots and velocity at different concentrations. At low concentrations of 5%, the size of spindle knots seems the same. At 6%, when the velocity is high (>175 mm/min), the size increases. While, at 7%, the size firstly increased and then decreased with the drawing-out velocity. This is an interesting investigation, which shows that the size (including both width and height) of the spindle knots varies to some extent. We can evaluate the water collection ability of BASs with different spindle knot sizes.

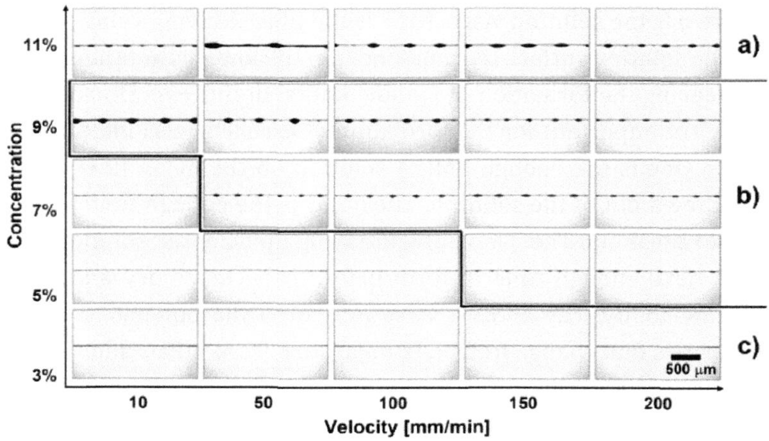

Figure 3.14 Optical images of BASs fabricated at different conditions by drawing the nylon fibers out of the PMMA/DMF solution.

3.2.1.2 Tiny water drops motion

Guided by the detailed mechanistic insights into directional water collection on spider silk gained in this study, Zheng and her team designed and fabricated artificial fibers that mimic the structural features of wetted rebuilt spider silk. Figure 3.15 shows the periodic spindle knots optical image of artificial spider silk. The spindle knots resemble those of the wetted spider silk and the periodicity of 394.6 ± 16.1 μm. The diameters of spindle knots and joints are 43.7 ± 5.4 μm and 13.5 ± 0.7 μm, respectively (Fig. 3.15b). They have the same microstructures with the wetted spider silk. The artificial joints have a porous structure mimicking aligned nanofibrils (Fig. 3.15c), and the artificial spindle knots exhibit random porous surface structures (Fig. 3.15d). The functional performance of the material is shown by the series of images in Figs. 3.15e–j. Figure 3.15e shows that small drops start to condense randomly on the artificial spider silk. When the drop volume increases (Fig. 3.15f), water drops move toward spindle knots from joints (Figs. 3.15g–j). These observations clearly illustrate that artificial spider silk not only mimics the structure of wetted rebuilt spider silk but also its capability of directional water collection.

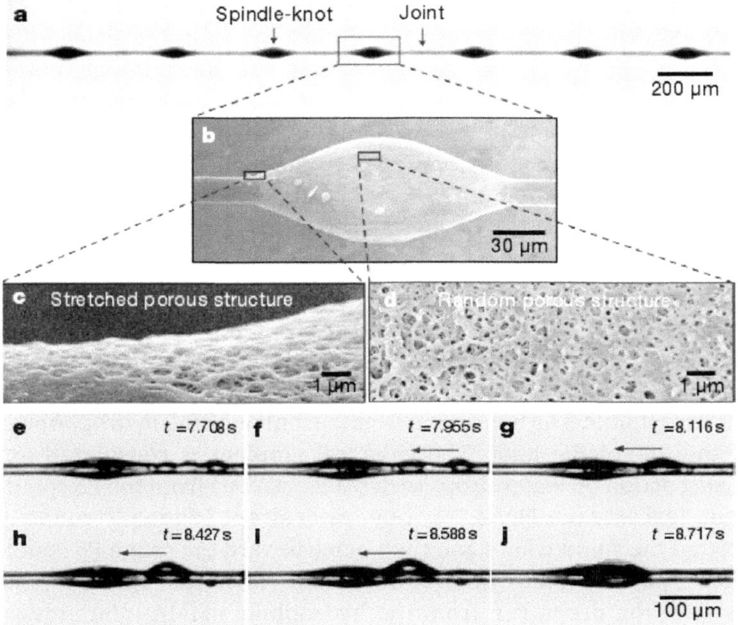

Figure 3.15 Artificial spider silk that mimics the structure and water collection capability of natural spider silk.

An example has been employed to show the tiny water motion on the BASs. The water drop can move toward or away from the spindle knots in the next examples. The prepared fibers with different kinds of spindle knots are tested in an environment with high RH of more than 95%. A charge coupled device (CCD) camera records the water drops motion images with time scale. Figure 3.16a shows a fiber with a rough PMMA spindle knot and a smooth PMMA joint. The tiny water drops firstly condensed randomly and then driven toward the spindle knot. Figure 3.16b also shows the case on the fiber with a rough hydrophilic polyvinyl acetate (PVAc) spindle knot. Quéré et al.[35] reported silicone oil drops movement toward the larger radius region on a smooth conical copper fiber. The curvature gradient along the fiber is considered as the driving force. However, tens of micrometers tiny drops are more difficult to drive. This is because that one driving force from curvature or chemical gradient is not enough to overcome the CA hysteresis. Daniel et al. tried to introduce more forces to drive water drops by using the energy released during coalescence on a smooth flat surface. In the case we discuss, smaller water drops are driven directionally at fast velocity on BASs. This case is a result of a more complicated mechanism. There is the roughness gradient along the fiber. Rough porous nanostructures are on the spindle knot and smooth structures are on the joints. Based on the principle of surface energy gradient, a water drop tends to be driven toward the higher surface energy or the more wettable region. As a result, the spindle knots are divided into two parts: one is covering PS with a lower surface energy than PMMA that is the other part. The water drops are driven away from the spindle knot. In addition, according to Wenzel's equation, $\cos\theta' = r\cos\theta$, where θ', θ, and r are, respectively, the apparent, intrinsic CAs and the roughness factor; roughness increases the surface wettability when it is hydrophilic and decreases that when it is hydrophobic. The nanostructures can enhance the hydrophobicity of the PS spindle knot. The chemical gradient is considered as a strong action on water drops to drive them away from the PS spindle knot. Just like the Fig. 3.16c, tiny water drops randomly condense around the spindle knot and then move toward the rough PS spindle knot from two sides of the joints. This is an unexpected result because the drops run from the hydrophilic PMMA joints toward the hydrophobic PS spindle knot. Figure 3.16d shows the similar behavior on the fiber with the rough PVDF spindle knot. What is interesting is that roughness cannot enhance chemical gradient to drive the drops toward the more wettable region (smooth PMMA joint). In this case, another factor should be taken into account to understand this unusual phenomenon. A drop must overcome the resistant force, because of CA hysteresis, to move, which is always opposite to the moving direction.

To get rid of the influence of roughness, the fibers are treated in solvent atmosphere to be smoother on the fiber surfaces. Figures 3.14e–h show the behavior of tiny water drops on these fibers. In Fig. 3.16g, the smooth hydrophobic PS spindle knots and smooth hydrophilic PMMA joints are investigated. Water drops are driven at 340.5–560.9 μm/s from the spindle knots to the joints. Figure 3.16h illustrates the similar behavior on the fiber with the smooth PVDF spindle knot. Compared to Figs. 3.14a and b with smooth surfaces, the water drops are driven toward the spindle knots.

This study indicates that three main factors control the tiny water directional driving. Firstly, chemical force is from the chemical gradient:

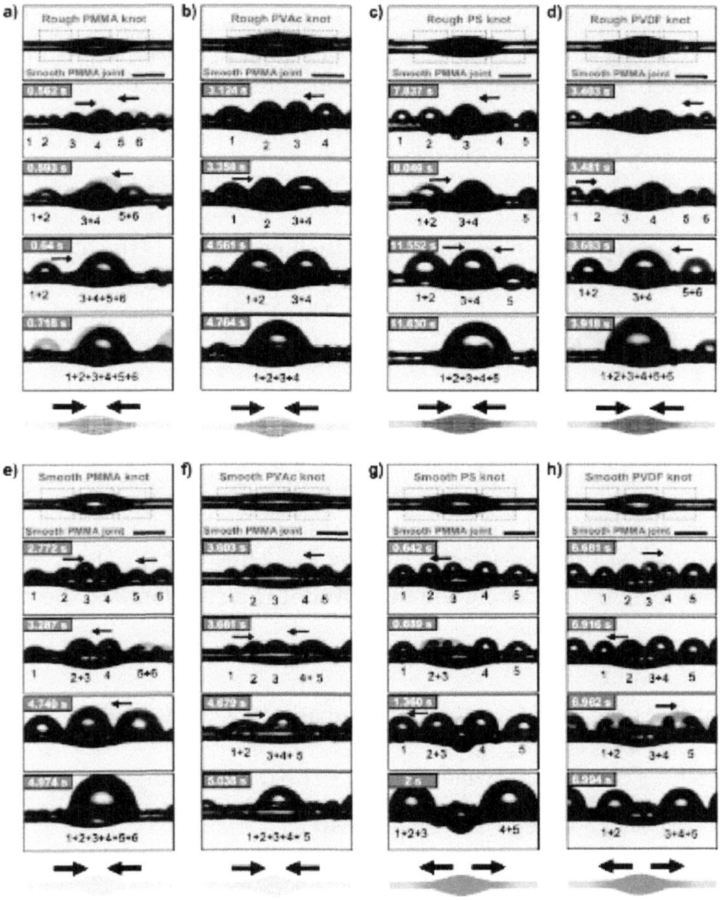

Figure 3.16 In situ optical observation of driving tiny water drops with controllable direction on artificial spider silks. Tiny water drops were driven toward the knots on the fibers with rough (a) PMMA, (b) PVAc, (c) PS, and (d) PVDF spindle knots. After the fibers were smoothened, drops were driven to different directions: toward the knots on the fibers with smooth (e) PMMA and (f) PVAc spindle knots or away from the knots on the fibers with smooth (g) PS and (h) PVDF spindle knots. These spindle knots and joint regions are indicated by blue and red rectangles, respectively. The drops were numbered to track their moving directions (indicated by black arrows). The bottom schemes illustrated fibers with rough or smooth spindle knots, where the moving direction of tiny water drops could also be seen in black arrows. Scale bar is 50 μm.

$$F_C \approx \pi R_0 \gamma_{water}(\cos\theta_J - \cos\theta_K) \tag{19}$$

where γ_{water} is the surface tension of water, R_0 is the radius of the coalesced bigger drop, θ_J is the local static CAs at the joint, and θ_K represents that at the spindle knot. Secondly, because of CA hysteresis, the hysteresis resistance is

$$F_H \approx \pi R_0 \gamma_{water}[(\cos\theta_{rJ} - \cos\theta_{aJ}) - (\cos\theta_{rK} - \cos\theta_{aK})] \tag{20}$$

where $\cos\theta_{rJ}$ is the receding CAs at the joint, $\cos\theta_{aJ}$ is the advancing CAs at the joint, and $\cos\theta_{rK}$ and $\cos\theta_{aK}$ are, respectively, the receding and advancing CAs at spindle knot. Thirdly, the curvature gradient induces the Laplace force

$$F_L \approx \gamma_{water}\left(\frac{1}{R_J'} - \frac{1}{R_K'}\right)\frac{\sin\alpha}{R_K - R_J}\frac{4}{3}\pi\left(\frac{R_0}{2}\right)^3 \tag{21}$$

where R_J and R_K are, respectively, the local radius of the spindle knot, R_J' and R_K' are the local contact lines curvatures of the drop at the two opposite sides along the fiber. And it is also related to the drop volume $\frac{4}{3}\pi\left(\frac{R_0}{2}\right)^3$. In summary, for a tiny water drop on a spindle knot fiber, the total force is described as follows:

$$F_{total} \sim \pi R_0 \gamma_{water}(\cos\theta_J - \cos\theta_K) - \pi R_0 \gamma_{water} \left[(\cos\theta_{rJ} - \cos\theta_{aJ}) - (\cos\theta_{rK} - \cos\theta_{aK})\right] - \gamma_{water}\left(\frac{1}{R_J'} - \frac{1}{R_K'}\right)\frac{\sin\alpha}{R_K - R_J}\frac{4}{3}\pi\left(\frac{R_0}{2}\right)^3 \tag{22}$$

Tiny water drops can be driven toward the spindle knots when $F_{total} < 0$. Water drops are driven away from the spindle knots when $F_{total} > 0$. In the experiments mentioned above, the size of the spindle knots is the same as the height of ≈34.9 μm and length of ≈128.5 μm. So F_L is a constant factor. The $\cos\theta_J$, $\cos\theta_{rJ}$ and $\cos\theta_{aJ}$ are also the constant factors due to joints of all the fibers with smooth PMMA. As a result, the static and dynamic CAs on the spindle knot determine the directional driving of drops. Namely, the value of F_{total} is determined by $\cos\theta_K$, $\cos\theta_{rK}$ and $\cos\theta_{aK}$.[41]

3.1.3 The Research of Three Phase Line

The geometric features of a fiber are significant to improve the drop behavior efficiency. Laplace pressure gradient can drive the drops movement toward the maximum diameter end of a conical fiber, or be propelled anisotropically on a mammalian hair fiber, or cover along the axis of a stripe-structured fiber. In our report, tiny water drops can move directionally toward the rough spindle knots on the multigradient-designed artificial spider silk. Here, we will consider another phenomenon of large hanging-drops on the fiber. After tiny water coalescence to form a larger hanging drop, it is attractive to study the force. The capillary adhesion and gravity of drop is balancing on the fiber. Hung et al.[65] had shown a simplified model based on dimensionality. They deduced the largest attachment force to be $2\gamma D$, where γ and D are, respectively, the liquid surface tension and the diameter of a uniform fiber. Quéré et al. assumed that the hanging force from a thin fiber was balanced by the force from two similar fibers radially to the drop at the same location. Quéré et al. predicted a larger hanging force of $2\pi\gamma D$ for a wetted fiber. In fact, the model shows that at the solid/liquid/gas three-phase contact region, there is local distortion of the fiber. So they could obtain a much higher V_{thr} value. We will introduce about the experimentally evaluated threshold volume (V_{thr}) before a drop separating away from a given fiber. The hanging drop has a shape nearly ellipsoidal configuration $\mathrm{V_{thr}} = 4\pi \mathrm{r^2{}_a r_b}/3$ where r_a and r_b are the width and length of the ellipsoidal drop, respectively. However, the uniform fibers show a larger V_{thr} value than predicted by Quéré et al. Figure 3.17 shows the V_{thr} of uniform fibers with diameters of 18 μm and 50 μm are, respectively, of ~2.03 ± 0.11 μL and of ~3.32 ± 0.08 μL. At room temperature, the γ is of 7.28 × 10^2 N/m. The value of V_{thr-18} is much more than $2\pi\gamma D$ and so is V_{thr-50}. We have introduced that the BAS fibers have a porous morphology. The V_{thr} of smooth nylon fiber can be investigated to analyze the effect of roughness. The research has found that the roughness shows a limited effect in enhancing V_{thr}. Moreover, we cannot explain difference between this study and previous ones by the wettability of the nylon fiber. In fact, considering a certain CA θ for the nylon fiber to water, the deviation should be more outstanding. Because the previous model provides a maximum capillary force of $2\pi\gamma D\cos\theta$, it is less than $2\pi\gamma D$ for a fiber with a CA θ.

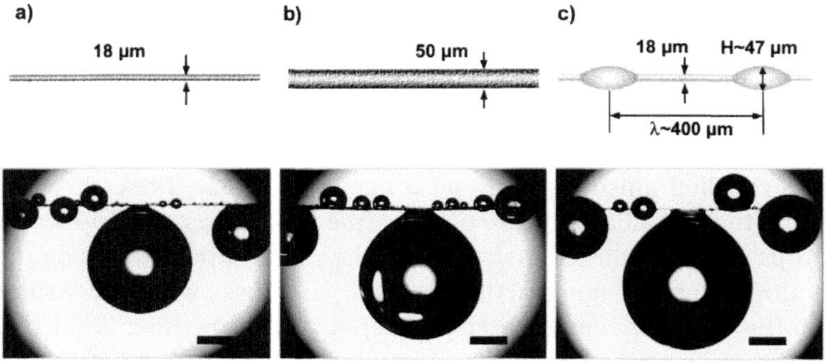

Figure 3.17 Hanging-drops on uniform fibers and a humped fiber. (a) Uniform fiber ($D \sim$ 18 µm), V_{thr} is 2.03 ± 0.11 µL. (b) Uniform fiber ($D \sim$50 µm), V_{thr} is 3.32 ± 0.08 µL. (c) Humped fiber (hump height H of ~47 µm and hump–hump distance λ of ~400 µm), V_{thr} is 4.38 ± 0.13 µL. Scale bars is 500 µm.

High-magnification optical microscopy is used to explore the origin of the observed unprecedented V_{thr}. The process of water condensation is observed on a uniform fiber. For a larger hanging water drop, it adheres to the lower surface of the fiber and does not surround it. The top of fiber is nonwetting. Due to Laplace pressure difference, if wetting the upper surface of the fiber, the tiny drops would rapidly be coalesced into the much larger hanging one. So like Fig. 3.18b, only the lower surface of the fiber is wetted by the hanging water drops. TCL is identified to extend along the fiber axially rather than being limited at the two ends of a wetted fiber. The length of TCL is much larger than $2\pi D$, which makes it possible to provide a capillary force more than $2\pi\gamma D$. To show more detail about the capillary action, d_u is used to denote the length of TCL from side view in Fig. 3.18b. With the growth of a hanging drop, the d_u is shortening. It makes the drop grow into a threshold volume, which is corresponding to a critical balance between gravity of a drop and the capillary force. The Laplace pressure at the solid/liquid interface can explain the growth of hanging drops. When the drops grow into a ellipsoidal shape, the Laplace pressure can be quantified $2\gamma/r_a - 2\rho g r_b$ in which ρ is the liquid density. The Laplace pressure at the interface will decrease with the drop growth, and the TCL is unstable in the length with the "dewetting" transition.

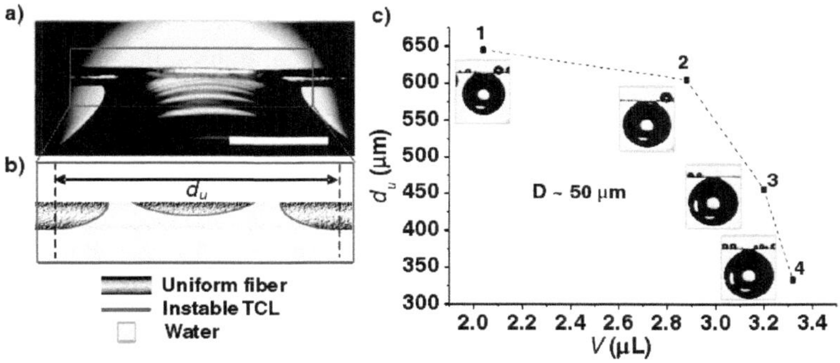

Figure 3.18 Instable TCL on uniform fiber during the growth of a hanging drop.

As known, large water drops can pin stably on thin spider capture silk in nature. The geometry of wetted spider silk plays an important role in the phenomenon and affects the solid/liquid adhesion. Inspired by wetted spider silk, a series of spindle knots and joints artificial fiber is fabricated to achieve an ability of water collection. Figure 3.17 reveals that the hump structures of thinner fibers can generate a stronger adhesion even than thicker ones without humps. It helps the surface geometry modification. Figure 3.19a shows the optical microscopy image of interface contact region on the BAS fibers. The upper surface of the fiber can be covered by some tiny water drops, which illustrates the upper surface is nonwetted by water drop. d_{h} is defined as the total length at the solid/liquid contact region. It is between the outer ends of two spindle knots in Fig. 3.19b. With the hanging drop growth, the TCL does not shrink, and d_{h} keeps almost constant (Fig. 3.19c). The capillary force can be quantified along the TCL as follows:

$$F = 2\int_{x_l}^{x_r} \gamma\cos\varphi(x)dx = 2\int_{x_l}^{x_l+d_h} \gamma\cos\varphi(x)dx \qquad (23)$$

in which x, x_{l}, and x_{r} are the coordinate variable along the fiber axially, the initial, and terminate sites for the integration at the outer ends of the left and right spindle knots, respectively, and $\varphi(x)$ is the local crossing angle between the vertical reference region and the liquid surface of the fiber. Compared to the uniform fiber (Fig. 3.18), the spindle knots on a BAS fiber can prevent the TCL to recede. Even though the Laplace pressure at the solid/liquid interface decreases, the d_h remains constant. It can promote the capillary adhesion.

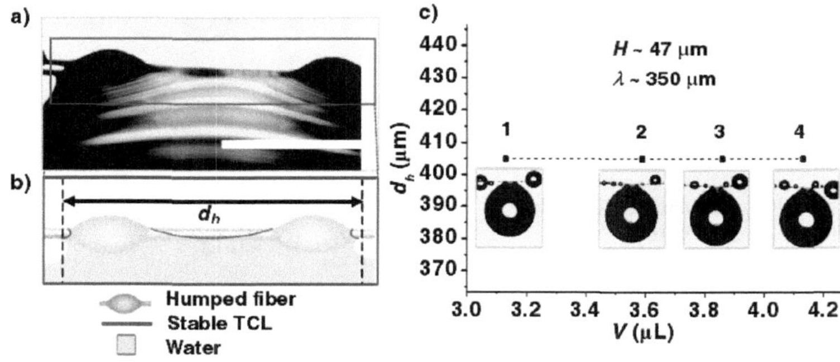

Figure 3.19 Stable TCL on artificial humped fiber during the growth of a hanging drop.

The basic geometrical features of spindle knots result in the two unique effects on the hanging action of humped fibers. One is the "slope" effect. Figure 3.20 shows that the lateral contact line tends to run away from the site "A" to "B" and to "C" along the spindle knot. Accordingly, because of the variation of surface slope the apparent receding angle θ_r of water reduces from $\theta_{r0} + \alpha \rightarrow \theta_{r0} \rightarrow \theta_{r0} - \alpha$. Here, the θ_{r0} and α are, respectively, the true receding angle of water on the surface and the semiapex angle of spindle knot. For a certain pendent drop with decreasing θ_r, it needs to activate the receding event at the contact line by a larger liquid surface distortion. So the slope variation provides a barrier for the contraction of TCL. The other is the "curvature" effect. Lorenceau and Quéré[35] studied that a liquid drop on a conical wire attempts to move toward the lower curvature location because of Laplace pressure gradient. The maximum diameter part of the spindle knot has a lower curvature than other region. It is reasonable that a water drop can move toward the hump. Figure 3.20 shows more details about the solid–liquid contact line receding from location "B" to "C." As a result, the curvature radius of the liquid surface at the solid/liquid contact line will reduce from $H/2$ to $D/2$. It results in the Laplace pressure increasing at the contact line and a larger CA. By the cooperating action of "slope" and "curvature" effects, the contact line is difficult to pass through "*B*" location on the spindle knot. The spindle knot can guarantee the relative stability of TCL and make sure a sufficient capillary force to capture drops. In addition, we have discussed the effect of roughness. The rough surface composed of porous structure may increase the ability of capturing drops to an extent. The pinning action of TCL can be enhanced by the roughness with the cooperation with the effects of curvature and slope of hump. This cooperation can lead to a larger volume of the extreme hanging-drop. There is not the recession barrier of contact line on a uniform fiber due to absence of a geometrically heterogeneous structure. So the TCL is instable and can easily slide under outer stimulation. When the Laplace pressure at the solid/liquid interface decreases, the TCL inevitably leads to capillary force to hang a limited number of water drops.

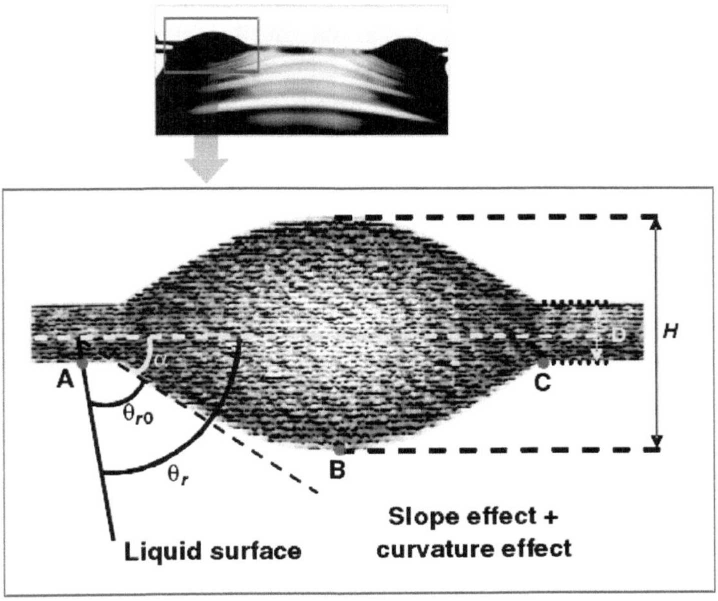

Figure 3.20 Illustration of the "slope" effect and "curvature" effect.

In summary, the study on BAS fibers open a wide insight into the interface adhesion on curved solid surfaces. And it provides a base to the design's one-dimensional functional materials by introducing special geometrically heterogeneous structures. These findings are important in many potential applications. For example, in the detection of nanobacteria, sample assays, single-molecule spectroscopy and microreactors, filtering aerosols, the transport of microdroplets, and water-harvesting projects.

3.2.2 Fluid-Coating Method

Learning from the nature, bio-inspired materials mimic the water-collection ability and their structural features of spider silks. The ability of water collection has drawn many people's attention in interdisciplinary fields. In our previous discussion, BAS fiber is fabricated with periodic spindle knots by a dip-coating method. And we have systematically investigated their water collection ability. However, the dip-coating method is limited because of its discontinuity. We cannot obtain a very long length of fiber using this method with a single "immersing-in/drawing-out" operation. So it requires us to find a new method to fabricate durable and inexpensive BAS fibers and make the bio-inspired material applied in large scale practical applications. Here, we introduce another method named fluid-coating method to continuously fabricate periodic spindle knots on nylon fibers on a large scale. The ability of directional water collection will be investigated in this section.

Figure 3.21a shows the schematic illustration of the fluid coating method for the large-scale fabrication of BAS fibers. In order to avoid gravity-induced liquid flow, a nylon fiber is fed horizontally through a polymer solution reservoir. The fiber is fixed with one end connected to a rolling motor to drag the fiber through the polymer solution reservoir. Two capillary tubes are used to guide the fiber. When the fiber is steadily drawn out of the reservoir at a given velocity, the polymer solution is coating onto fiber surface at the end of the capillary tube. A CCD camera is used to record the whole fabrication process. Figures 3.21b and c show as-prepared bio-inspired fibers with periodic spindle knots on a large scale. The spindle knots with pure white color can be seen easily in Fig. 3.21c. The SEM images show the size of a typical spindle knot in Fig. 3.21e. The spindle knot has random and stretched porous structure like spider silk, respective, on the middle and two side regions shown in magnified SEM images. These investigations demonstrate that using fluid coating method can fabricate the BAS fibers with the structure similar to the spider silks.

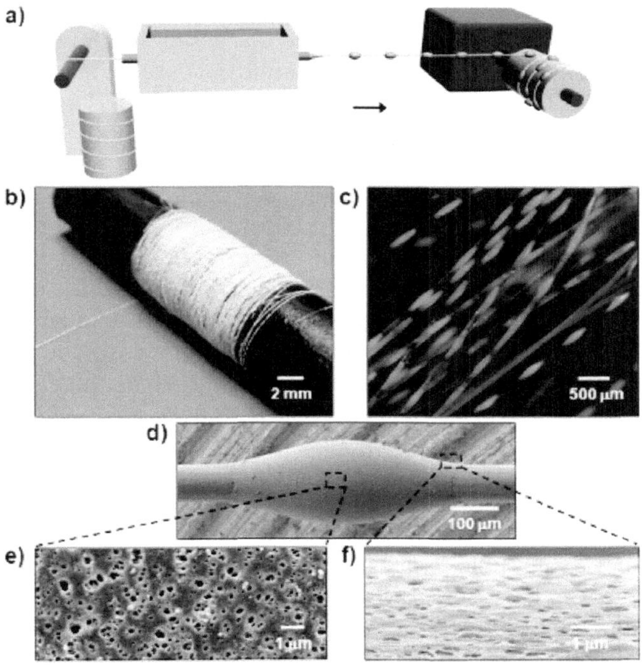

Figure 3.21 (a) Schematic illustration of the fluid coating method used for the large-scale fabrication of bio-inspired fibers. (b & c) Optical images of the as-prepared bio-inspired fibers with periodic spindle knots. (d–f) SEM image of a single spindle knot.

The fluid coating on fiber has been studied for a long time as a very common method. Its applications range the improvement of the mechanical properties of fibers to the corrosion protection of metal wire, even includes simply make candles. Using this method, we should consider how to control the film thickness under this fabrication conditions. Because of the equilibrium between the viscous forces, the dynamic meniscus makes the liquid move with the fiber. The force of the surface tension tends to drive the liquid back to the solution reservoir. The ratio of these two opposing forces largely affects the film thickness. The capillary number (C_a) can be described as follows: $C_a = \eta V/\gamma$, where η, V and γ are, respectively, the solution viscosity, the fiber drawing velocity, and the surface tension of the solution. When the polymer solution is given, the solution viscosity and surface tension of the solution cannot be changed. The coating behaviors base on the different drawing velocities to change. Here, three kinds of coating behaviors are discussed for the fabrication conditions. At low velocities, the dynamic meniscus is stable and the film thickness generally obeys the Landau, Levich, Derjaguin theory. This is called as the "viscocapillary region." For both PMMA and PVDF solution, the coating behavior is recorded at the end of the capillary tube at a very slow velocity. We think this behavior as the first kind of coating, in which the solution film cannot generate an obvious spindle knot after solvent evaporation due to the too thin film. When increasing the drawing velocity, the thickness of coated film also increases accordingly. The film is thick enough so that periodic spindle knots can be formed on the fibers by the Rayleigh instability. This is defined as the second kind of coating. The third kind of coating represents very high velocities. It is important to consider the effect of inertia. We call the regime as the "viscoinertial regime." In this condition, it is unstable for the dynamic meniscus. In the setup we presented above, the length of reservoir is 35 mm, and the inner diameter of capillary tube on its wall is about 400 μm. It is possible to observe the process of droplets expulsion from the capillary tube. After solvent evaporation, the polymer droplets become large spindle knots on the fiber. And the periodicity is as good as the fabricating method of dip-coating method by Rayleigh instability. Some small polymer droplets cannot form obvious spindle knots since their size is largely reduced after solvent evaporation.

The BAS silk fibers by fluid-coating method have been shown the directional water-collection ability. A CCD camera has observed the whole process that bio-inspired fibers fix it horizontally under a vertical fog flow. Figure 3.22 shows a typical PMMA spindle knot. Figure 3.22b shows the tiny water drops at the first time condense randomly around the spindle knot. The arrows in Figs. 3.22c and d show the water droplets collection directional toward the spindle knot. A large water droplet is formed finally (see in Fig. 3.22e). This investigation demonstrates that the BAS fibers have the water directional collection ability. The cooperation of two driving forces results in this amazing ability. One is the Laplace pressure gradient, which is arising from the curvature difference of the spindle knot shape. The other one is the surface energy gradient resulting from the difference in surface roughness. It is reported that under the same conditions nature spider silk can collect water less than BAS fiber. Because of the differences of fiber radius, surface structure, and chemical composition between the bio-inspired fibers and natural spider silks, the shallow comparison is worth nothing. So in order to improve the water collection efficiency of BAS fibers, the important factors should be optimized.

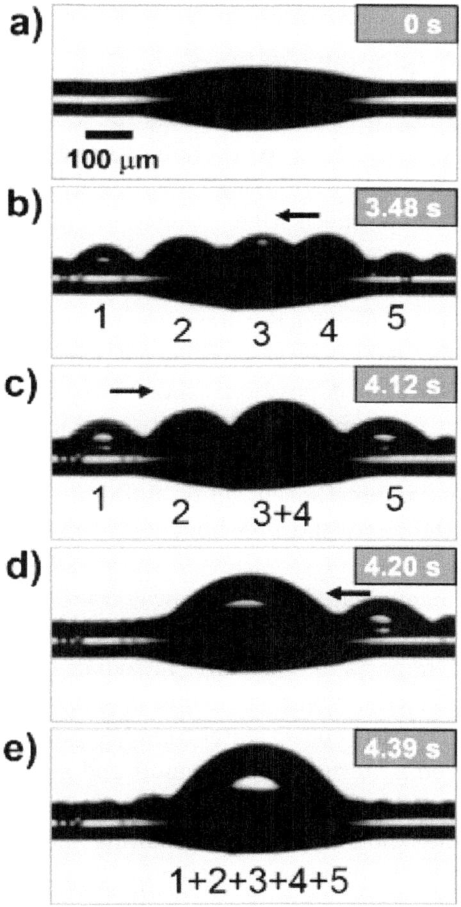

Figure 3.22 In-situ observation of the directional water-collection process on a bio-inspired fiber with a PMMA spindle knot.

In conclusion, the fluid-coating method can be used to fabricate bio-inspired fibers with spindle knots and joints on a large scale. We discuss the solution parameters and fiber-drawing velocities of formation of the spindle knots in detail. Importantly, the water directional collection ability is shown in this section. These observations are important for large scale fabrication with a low cost. These durable BAS fibers can be applied in water collection, directional transport of liquid droplets, and so on.

3.2.3 Electrospinning Methods

Wenzel and CB theories have testified that the roughness on surface is essential to enhance both hydrophobicity and hydrophilicity. In practice, we categorize the methods of superhydrophobic surface preparation into two aspects. One is called "top-down" method, and the other is "bottom-up." Top-down approaches include template-based and lithographic techniques, and plasma treatment of the surfaces. Bottom-up approaches involve mostly self-assembly and self-organization. For examples, chemical deposition, hydrogen bonding, layer-by-layer deposition, colloidal assemblies, and so on. Electrospinning is normally used for the preparation of polymer nanofibers. Here, we will introduce a method of electrospinning based on the combination of both bottom-up and top-down approaches to fabricate the BAS fiber with the water directional collection ability.[42]

We know, some creatures live far away from water. So water is one of the most important substances in their life. Water collection becomes very important. We have introduced some creatures having water collection abilities, such as desert beetles and spider silks. Desert beetles can collect water from foggy air because of their micrometer-sized patterns of hydrophobic and hydrophilic regions on their backs. Spider silks have special structures consisting of periodic spindle knots and joints resulting in water collection properties. Water drops move from the joints parts to the spindle knots. Natural creatures evolved superior performances to adapt to the environment. Learning from the nature, new functional materials are designed. We introduced dip-coating and fluid-coating methods to fabricated artificial spider silks to realize the water collection from humid air. Now, we will give another method, electrospining method. Electrospining is using electrohydrodynamics to stretch a viscous liquid into fibers. It has become an effective avenue to generate microscopic materials. It has reported that electrospinning can fabricate beaded fibers with a structure of periodic knots similar to the shape of natural spider silk.

Tian et al.[39] reported that using coaxial electrospinning (co-ESP) to fabricate various microscopic core-shell or tubular fibers from various materials. Traditional co-ESP chooses a highly viscous solution as the outer solution to form a uniform liquid shell restricting the inner solution. After the liquid thread solidifies, the hump structured thin fibers are formed. In these processes, the Rayleigh instability of the shell solution is overwhelmed by the viscoelasticity. If the viscoelasticity cannot overwhelm the Rayleigh instability, the liquid thread will fluctuate into knotted fiber or even break up into core-shell particles. In the earlier studies, the inhibition of Rayleigh instability in the co-ESP process is obtained more attention. In our mentioned method, it is different from the conventional coaxial electrospinning. It adapts the cooperation of the electrospraying of an outer fluid and electrospinning of an inner fluid. A sprayable outer fluid imprints a series of heterogeneous beads, and the spinnable inner fluid forms the centric fiber. The components of the inner and outer fluids should be controlled to form the bead-on-string heterostructured fibers with tunable compositions. Take an example, the outer fluid can choose the solution with 20% (w/v) poly(ethylene glycol) (PEG)/DMF, and the inner fluid is solution with 35% (w/v) polystyrene (PS)/DMF, and the distance between the needles and the aluminum foil is 22 cm with 19 V of applied voltage. The aluminum foil is used to accept the samples after coaxial electrohydrodynamic jetting process.

Figure 3.23 shows the ESEM image of capture silk and the illustration of electrospinning/electrospraying approach. In Fig. 3.21c, the red color is the spinnable inner fluid with high viscosity and the green color is the sprayable outer fluid with low viscosity. They combine and work through the inner and outer needle, respectively. Then, bead-on-string heterostructured fibers are collected on the aluminum foil. Figure 3.23d shows the typical SEM image of bead-on-string heterostructured fibers. The diameter of the joints ranges from a few micrometers to 10 μm. That of spindle knots is from a few micrometers to more than 20 μm. The energy-dispersive X-ray spectroscopy has verified that the PEG spindle knots are covered on the PS fibers by the main element carbon on both spindle knots and joints parts and the oxygen presenting only in the spindle knots parts. These investigations prove that the spindle knot is composed of pure PEG and the component of the joint part is only PS.

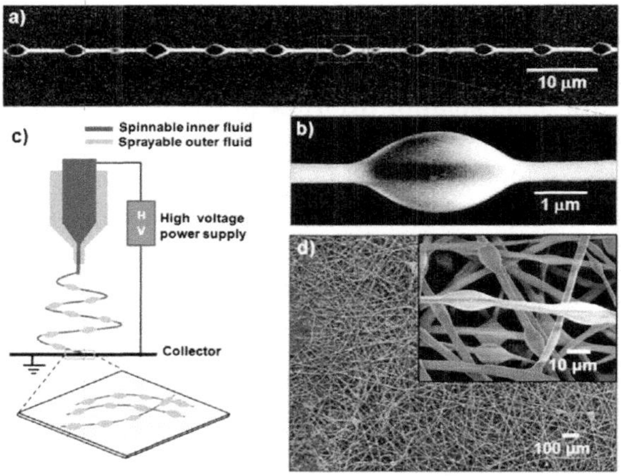

Figure 3.23 (a) ESEM image of capture silk of an ecribellate spider with regularly distributed glue droplet on it. (b) Magnified ESEM image of a glue-droplet on the capture silk. (c) Illustration of combined electrospinning/electrospraying approach. (d) SEM image of the spindle knots fibers, accompanied by a magnified image (inset).

For the preparation of these bead-on-string heterostructured fibers process, two fluids of inner/outer cannot be miscible. The electrohydrodynamic process can prevent them from mixing at a short-time. It is different from the typical coaxial electrospinning. This experiment uses a spinnable outer fluid with low viscosity, which is hard to envelope the inner fluid if using an immiscible solvent system. So we cannot use two fluids easily separating in the jetting process to fabricate the bead-on-string heterostructured fibers. A comparison experiment has tested the two immiscible inner/outer solvent systems: toluene/water and butanone/water, which both fail to form the bead-on-string heterostructured fibers.

The other effect factor to formation of bead-on-string heterostructured fibers is the flow rate of the inner fluid. If it is too high, the spindle knots structure decreases evidently or even disappears. Whereas if it is too low, the fluid is difficult to be shaped and drips from the nozzle. Increasing the inner fluid makes the decrease of the relative content of outer fluid, which will result in the decreasing of spindle knots structure. In this design, the surface energy difference is considered. PEG has a high surface energy, and PS has a low one. They are chosen as the components of the outer and inner fluid, respectively. As described above, these two fluids should be miscible. However, we can manipulate the interface action by choosing appropriate solute components. The viscous drag force exerted by the inner fluid can be mitigated by the outer polymer with higher surface energy and inner polymer with lower surface energy. During the electrohydrodynamic jetting process, the outer polymer favors the spontaneous Rayleigh break-up. In addition, the higher surface energy component can prevent the lower energy fluid wetting and adhesion during the solvent evaporation. These effects favor the formation of perfect bead-on-string heterostructures.

Figure 3.24 shows the response of bead-on-string heterostructured fibers to environmental change due to their alternating hydrophilic and hydrophobic surface characteristics. The ESEM images illustrate the segmented swelling behavior. At the RH of about 55%, the diameter of the spindle knot is about 7.2 μm, and it increases to 9.5 μm when the RH increased to more than 100%. However, the diameter of the middle part is remaining at 2.2 μm during the whole process. The swelling behavior is quite similar to that of the capture silk of ecribellate spiders in a humid environment.

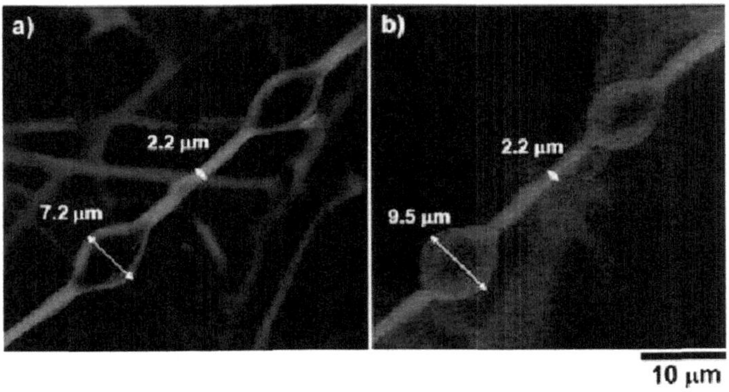

Figure 3.24 Segmented swelling behavior of a bead-on-string heterostructured fiber (BSHF) observed inside the ESEM (a) at 55% humidity, (b) >100% humidity.

This is a facile and powerful method to prepare novel BAS fibers. It opens the field of electrohydrodynamic jetting to a much wider range of surface chemistries and applications. The method is simple and it offers a possibility to quickly fabricate low-dimensional heterostructured materials on a large scale. Thanks to the electrohydrodynamic technique, this method can construct heterostructured materials by alternating surface configurations using a wide range of combinable component materials, including inorganic materials, polymers, and hybrid materials with various functional groups such as electronic, hydrophobic, optical, and hydrophilic chemical components. These kinds of BAS fibers are important in integrated functional devices or other many important properties and applications.

3.3 Promising Applications

3.3.1 Control Motion of Liquid Droplet

From the above section, we find that functional fibers with unique wettability and hump structure can be achieved in many ways. Zheng et al. further investigate the detailed mechanism of driving tiny water droplets on bio-inspired artificial fibers by designing the surface structures and chemical compositions of spindle knots. For example, the surface energy differences can be brought about by choosing different polymers including PMMA, poly(vinyl acetate), PVDF, and PS. The moving behavior of condensed water droplets is investigated. The results illustrate that tiny water drops can be driven in a controllable direction toward or away from the spindle knot. When the condensed water drops on the spindle knot and joint start to grow big enough, the coalescence will appear. Because of the unbalanced forces acting on the two opposite sides of the drops' contact lines, the coalesced bigger droplet can move toward or away from the spindle knot. In fact, there are three factors to consider, including curvature, chemical, and roughness gradients on the bio-inspired fibers. The three factors consist of the driving forces to realize the different directions and magnitudes.

3.3.2 High-Efficiency Water Collection

Water shortage is a world question. We provide a feasible route by learning from the spider silks to obtain fresh water by collecting water from fog air, especially in dry and fog-laden areas. The BAS fibers with periodic spindle knots have been investigated about the water collection ability above. The spindle knot geometry plays an important role in the process. It is very significant to apply these bio-inspired fibers as the water collection materials. By varying the size of spindle knots using dip-coating method or fluid-coating method, we can fabricate a series of the bio-inspired fibers. The solution viscosity, solution surface tension, and fiber drawing velocity are the crucial three factors. The investigations show that optimizing the conditions can obtain the geometry (the appropriate width and height of spindle knots) to get the largest volume of water drops. The experiment has shown that the bio-inspired fibers with larger spindle knot collect more water than that with smaller ones. The fibers fabricated by microfluidic technology have got the same results. The spindle knot quickly transport the condensed water drops away, and make the original condensation sites free. It makes BAS fibers with periodic spindle knots structure more efficient at collecting water from fog than the fibers with uniform diameter.

3.3.3 Enhanced Solid–Liquid Adhesion

The surface wettability has an important aspect, solid–liquid adhesion. Especially, adjusting the adhesion between the fiber surface and liquid drop is very important to optimize the water collecting efficiency of fiber-based devices. We have demonstrated that the BAS fibers with periodic spindle knots have much stronger adhesion to liquid droplets than smooth fibers. The above section has shown that larger water drops can hang on the bio-inspired fibers. Due to the unusual TCL on the fiber surface with spindle knot structure, the fiber has a stable structure with "slope" and "curvature" effects. Engineering special and geometrically heterogeneous structures onto the fiber surface can enhance the solid–liquid adhesion.

3.3.4 Environmentally Responsive Fiber

Ecribellate spiders can spin a type of capture silk with periodic glue droplets along the main fiber. Inspired by the spinning characteristics and its geometric structure, using electrohydrodynamic method fabricates a kind of functional heterogeneous fibers with hump structure. The as-prepared fiber is heterogeneous with PEG spindle knots and PS middle parts. These bio-inspired fibers have the function to respond to the environmental humidity. ESEM image has shown the diameter change with the RH from 55% to about 100%. But the middle part does not swell. Such functional fibers with heterogeneous structures are capable of selectively responding to environmental stimuli. Maybe we can find applications in highly integrated functional devices.

Some functional fiber can be fabricated by methods such as dip coating, fluid coating, and electrodynamically. We have summarized the designing strategies and applications of bio-inspired functional fibers with unique wettability. Other designed works are going on. We may stress both aspects of structural features and chemical components onto the fiber surface to obtain novel functional fibers. For example, we introduce some light, pH, or thermal stimuli molecules modified on the fiber surfaces to control the moving direction of water droplets. We investigate the novel fabricating methods and techniques to obtain asymmetric structures or wettability gradient onto the spindle knot so as to achieve the unidirectional motion of water droplets. By designing the gradients, we will achieve fast and high-efficiency liquid transportation along the fiber. In addition, we can apply bio-inspired functional fibers with unique wettability in many fields, such as water collection, smart catalysis, filtration, and sensing. The water-collecting efficiency can be increased by directionally transporting small water droplets together to coalesce into large droplets. Smart catalysis can expedite the transporting speed of water drop toward or away from the spindle knot. The tiny liquid drops on the bio-inspired functional fiber can capture small amount of materials in the air (e.g., particles). Then by directionally collecting the liquid droplets toward the spindle knot further concentrates or filtrates the small materials.[43]

References

1. Gao XF, Jiang L (2004) Water-repellent legs of water striders, *Nature*, **432**(4), 36.
2. Lukas E, Andrew S, Thomas S (2011) Decoding the secrets of spider silk, *Mater Today*, **14**(3), 80–86.
3. Vollrath F, Porter D (2006) Spider silk as archetypal protein elastomer, *Soft Matter*, **2**(5), 377–385.

4. Nathan B, Emin O, Stephanie M, Jason PC, Paul KH, Cheryl YH, Dmitrii EM, Helen GH (2003) Molecular nanosprings in spider capture-silk threads, *Nat Mater*, **2**(4), 278–283.

5. Opell BD, Bond JE (2000) Extensibilities of cribellar and adhesive capture threads of orb-weaving spiders: Testing punctuated and associative explanations of character evolution, *Biol J Linn Soc*, **70**, 107–120.

6. Hawthorn AC, Opell BD (2003) van der Waals and hygroscopic forces of adhesion generated by spider capture threads, *J Exp Biol*, **206**(22), 3905–3911.

7. Peter HM (1984) The spinning apparatus of Uloboridae in relation to the structure and construxtion of capture threads (Arachnida, Araneida), *Zoomorphology*, **104**(2), 96–104.

8. Van Beek JD, Hess S, Vollrath F, Meier BH (2002) The molecular structure of spider dragline silk: Folding and orientation of the protein backbone, *PNAS*, **99**(16), 10266–10271.

9. Riekel C, Vollrath F (2001) Spider silk fiber extrusion: Combined wide and small-angle X-ray microdiffraction experiments, *Int J Biol Macromol*, **29**(3), 203–210.

10. Yi L, Zhengzhong S, Fritz V (2005) Relationships between supercontraction and mechanical properties of spider silk, *Nat Mater*, **4**, 901–905.

11. Holland C, Terry AE, Porter D, Vollrath F (2006) Comparing the rheology of native spider and silkworm spinning dope, *Nat Mater*, **5**, 870–874.

12. Porter D, Vollrath F (2007) Nanoscale toughness of spider silk, *Nanotoday*, **2**(3), 6.

13. Fiorenzo GO, David LK (2010) New opportunities for an ancient material, *Science*, **329**, 528–531.

14. Kinahan ME, Emmanouela F, et al. (2011) Tunable silk: Using microfluidics to fabricate silk fibers with controllable properties, *Biomacromolecules*, **12**(5), 1504–1511.

15. Edward K, Gi SJ, Yoon YC, et al. (2011) Digitally tunable physicochemical coding of material composition and topography in continuous microfibers, *Nat Mater*, **10**, 877–883.

16. Vasav S, Todd AB, Ali D (2010) Viscoelastic solids explain spider web stikiness, *Nat Commun*, **1**, 19.

17. Li X, Tian J, Shen W (2009) Thread as a versatile material for low-cost microfluidic diagnostics, *ACS Appl Mater Interfaces*, **2**(1), 1–6.

18. Zheng Y, Bai H, Huang Z, et al. (2010) Directional water collection on wetted spider silk, *Nature*, **463**, 640–643.
19. Lorraine HL, Edmonds DT, Vollrath F (1995) Structural engineering of an orb-spider's web, *Nature*, **373**, 146–148.
20. Lazaris A, Arcidiacono S, Huang Y, et al. (2002) Spider silk fibers spun from soluble recombinant silk produced in mammalian cells, Science, **295**, 472–476.
21. Lee SM, Pippel E, Gosele U, et al. (2009) Greatly increased toughness of infiltrated spider silk, *Science*, **324**, 488–492.
22. Jin HJ, Kaplan DL (2003) Mechanism of silk processing in insects and spiders, *Nature*, **424**, 1057–1061.
23. van Beek JD, Beaulieu L, Schafer H, Demura M, Asakura T, Meier BH (2000) Solid-state NMR determination of the secondary structure of Samia Cynthia ricini silk, *Nature*, **405**, 1077–1079.
24. Thiel BL, Viney C (1996) Beta sheets and spider silk, *Science*, **273**(5281), 1480–1481.
25. Yoshimitsu Z, Nakajima A, Watanabe T, Hashimoto K (2002) Effects of surface structure on the hydrophobicity and sliding behavior of water droplets, *Langmuir*, **18**, 5818–5822.
26. Fang G, Li W, Wang X, Qiao G (2008) Droplet motion on designed microtextured superhydrophobic surfaces with tunable wettability, *Langmuir*, **24**, 11651–11660.
27. Feng L, Li S, Li Y, et al. (2002) Super-hydrophobic surfaces: From natural to artificial, *Adv Mater*, **14**(24), 1857–1860.
28. Barthlott W, Neinhuis C (1997) Purity of the sacred lotus, or escape from contamination in biological surfaces, *Planta*, **202**(1), 1–8.
29. Parker AR, Lawrence CR (2001) Water capture by a desert beetle, *Nature*, **414**, 33–34.
30. Lee W, Jin MK, Yoo WC, Lee JK (2004) Nanostructuring of a polymeric substrate with well-defined nanometer-scale topography and tailored surface wettability, *Langmuir*, **20**(18), 7665–7669.
31. Liu M, Wang S, Wei Z, Song Y, Jiang L (2009) Bioinspired design of a superoleophobic and low adhesice water/solid interface, *Adv Mater*, **21**(6), 665–669.
32. Zheng Y, Gao X, Jiang L (2007) Directional adhesion of superhydrophobic butterfly wings, *Soft Matter*, **3**, 178–182.
33. Daniel S, Chaudhury MK, Chen JC (2001) Fast drop movements resulting from the phase change on a gradient surface. *Science*, **291**, 633–636.

34. Daniel S, Sircar S, Gliem J, Chaudhury MK (2004) Ratcheting motion of liquid drops on gradient surfaces, *Langmuir*, **20**(10), 4085–4092.

35. Lorenceau E', Que're' D (2004) Drops on a conical wire, *J Fluid Mech*, **510**, 29–45.

36. Huang Z, Chen Y, Zheng Y, Jiang L (2011) Capillary adhesion of wetted cribellate spider capture silks for larger pearly hanging-drops, *Soft Mater*, **7**, 9468–9473.

37. Yang J, Yang Z, Chen C, Yao D (2008) Conversion of surface energy and manipulation of a single droplet across micropatterned surfaces, *Langmuir*, **24**, 9889–9897.

38. Bai H, Ju J, Sun R, Chen Y, Zheng Y, Jiang L (2011) Controlled fabrication and water collection ability of bioinspired artificial spider silks, *Adv Mater*, **32**, 3708–3711.

39. Tian X, Chen Y, Zheng Y, Bai H, Jiang L (2011) Controlling water capture of bioinspired fibers with hump structures, *Adv Mater*, **46**, 5486–5491.

40. Hou Y, Chen Y, Xue Y, Wang L, Zheng Y, Jiang L (2012) Water collection behavior and hanging ability of bioinspired fiber, *Langmuir*, **28**, 4737–4743.

41. Bai H, Tian X, Zheng Y, Ju J, Zhao Y, Jiang L (2010) Direction controlled driving of tiny water drops on bioinspired artificial spider silks, *Adv Mater*, **22**, 5521–5525.

42. Li XM, Reinhoudt D, Crego-Calama M (2007) What do we need for a superhydrophobic surface? A review on the recent progress in the preparation of superhydrophobic surfaces, *Chem Soc Rev*, **36**, 1350–1368.

43. Bai H, Ju J, Zheng Y, Zhao Y, Jiang L (2012) Functional fibers with unique wettability inspired by spider silks, *Adv Mater*, **24**, 2786–2791.

44. Johnson RE, Dettre RH (1964) Contact angle hysteresis. III. Study of an idealized heterogeneous surface, *J. Phys. Chem.*, **68** (7), 1744–1750.

45. Cassie ABD, Baxter S (1944) Wettability of porpous surface, *Trans. Faraday Soc.*, **40**, 546–551.

46. Genzer J, Efimenko K (2000) Creating long-lived superhydrophobic polymer surfaces through mechanically assembled monolayers, *Science*, **290**, 5499–2130.

47. Sandre O, Gorre-Talini L, Ajadar A, Prost J, Siberzan P (1999) Moving droplets on asymmetrically structured surfaces, *Phys. Rev. E.*, **60**(3), 2964–2972.

48. Petrie RJ, Bailey T, Gorman CB, Genzer J (2004) Fast directed motion of "fakir" droplets, *Langmuir*, **20**(23), 9893–9896.

49. Shastry A, Case MJ, Bohringer KF (2006) Directing droplets using microstructured surfaces, *Langmuir,* **22**(14), 6161–6167.

50. Yang JT, Chen JC, Huang KJ, Yeh JA (2006) Droplet manipulation on a hydrophobic textured surface with roughened patterns, *J. Microelectromech. Syst.,* **15**(3), 697–707.

51. Bouasse H (1924) Capillarite, phenomenes superficiels, Delagrave, Paris.

52. Weisloge MM (1997) Steady spontaneous capillary flow in partially coated tubes, *AIChE J,* **43**, 645–665.

53. Bain C, Burnett-Hall G, Montgomerie R (1994) Rapid motion of liquid drops, *Nature,* **372**, 414–415.

54. Domigues dos santos F, Ondarcuhu T (1995) Free running droplets, *Phys. Rev. Lett.,* **75**, 2972–2976.

55. Bico J, Quéré D (2002) Self-propelling slugs, *J. Fluid Mech.,* **467**, 101–127.

56. Brochard F (1989) Motions of droplets on solid surfaces induced by chemical or thermal gradients, *Langmuir,* **5**, 432–438.

57. Smith MK (1995) Thermocapillary migration of a two-dimensional liquid droplet on a solid surface, *J. Fluid Mech.,* **294**, 209–230.

58. Mazouchi A, Homsy GM (2000) Thermocapillary migration of long bubbles in cylindrical capillary tubes, *Phys. Fluids,* **12**, 542–549.

59. Carroll BJ (1989) Droplet formation and contact angles of liquids on mammalian hair fibres, J. Chem. Soc. Faraday Trans., 185, 3853–3860.

60. Chaudhury MK, Whitesides GM (1992) How to make water run uphill, *Science,* **256**(5063), 1539–1541.

61. Grunze (1999) Driven liquids, *Science,* **283**, 41–42.

62. Darhuber AA, Valentino JP (2003) Thermocapillary actuation of droplets on chemically patterned surfaces by programmable microheater arrays, *J. Microelectromech. Syst.,* **12**, 873–879.

63. Yasuda T, Suzuki K, Shimoyama I (2003) Automatic transportation of a droplet on a wettability gradient surface, *Seventh International Conference on Miniaturized Chemical and Biochemical Analysis Systems,* 1129–1132.

64. Ichimura K, Oh SK, Nakagawa M (2000) Light-driven motion of liquids on a photoresponsive surface, *Science,* **288**, 1624–1626.

65. Hung LS, Yao SC (1999) Experimental investigation of the impaction of water droplets on cylindrical objects, *Int. J. Multiphase Flow,* **25**(8), 1545–1559.

Chapter 4

Beetle Back Effect: Heterogeneous Wetting Micro- and Nanostructure Patterns

4.1 Fog Collection (and Mechanism)

We have discussed that nature is an expert and has abilities of designing elaborate micro- and nanostructures, such as spider silks. Parker et al. reported that Namib Desert beetle could extract fog droplets from moist breeze efficiently, which is benefiting from the hydrophobic/hydrophilic patterned regions on its back.[1]

The Namib Desert in southern Africa experiences extreme daytime temperatures, high wind and early-morning fog,[2–3] but the rainfall is so low and variable that it can be considered negligible here. The tenebrionid beetle Stenocata (Fig. 4.1a) shows that its body collects water toward the wind. Water drops grow on the beetle's bumps and roll down to his mouth through the valley regions. Figure 4.1a shows that a near-random array of bumps (each about 0.5 mm in diameter) exists on the back of desert beetles on a macroscopic scale. Seen from the microscopic level, the valleys are covered by a microstructure coated in wax, whereas the top of these bumps are smooth. The microstructure is composed of flattened hemispheres with diameter of ~10 μm. The flattened hemispheres are arranged in a regular hexagonal array to achieve a

Bio-Inspired Wettability Surfaces: Developments in Micro- and Nanostructures
Yongmei Zheng

ISBN 978-981-4463-60-7 (Hardcover), 978-981-4463-61-4 (eBook)
www.panstanford.com

superhydrophobic system similar with the microstructure of lotus leaf.[1] Three main steps are involved in the process of extracting fog water droplets from air to the solid surface, that is, "deposition," "collection," and "transportation." About the first step "deposition," because the fog water droplets encountered by *Stenocara* are much finer (1–40 μm diameter) than raindrops and, without a unique controlling mechanism, would quickly be lost to the heat and winds of the desert. The tiny fog water droplets will keep growing after they get pinned on the top of bump, and the fog water droplets striking the hydrophobic valleys can also be collected, they may bounce or be blown to a hydrophilic region. This is the second step "collection." Each attached droplet eventually reaches a size at which its contact area covers the entire hydrophilic island. Beyond this size, the ratio of its mass to its surface contact area increases rapidly until the capillary force that attaches it to the surface is overcome (this force is dictated by the area of the hydrophilic island). At this point, the droplet detaches and rolls down the tilted beetle's surface, guided by the slight purchase afforded by other peaks along its path. This is the third step "transportation."

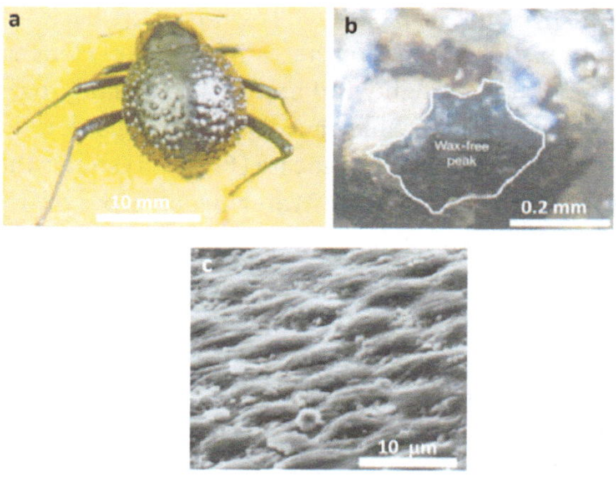

Figure 4.1 The water-capturing surface of the fused overwings (elytra) of the desert beetle *Stenocara* sp.

The efficiency of this water collecting system is determined by the efficiency of those three steps. If there is more fog water droplets extracted to the solid surface and transported away in unit time, of course the high efficiency is realized. So, the growing water droplets roll away from hydrophilic regions timely, a new circle of collecting would begin immediately. As described above, because the bumps can capture fog water droplets efficiently with its hydrophilic top, and the valleys can deliver the growing droplets quickly, the desert beetle's back indicates a very efficient water-collecting ability.

4.2 Artificial Pattern and Functions

4.2.1 Hydrophilic Bumps on (Super) hydrophobic Wax Films

Parker et al. mimicked the structure by embedding 0.6 mm glass spheres into microscope slides, which is coated with warm wax. In one model, they arranged the spheres in a square array with ~0.6 mm × 0.6 mm; while in another model, an average spacing of 0.5 mm arranged randomly with spheres. To compare two models, they used uniformly hydrophilic surface with bare glass and a uniformly hydrophobic surface with smooth wax. Ultrasonic agitation in isopropyl alcohol cleaned it for 15 min.

Four samples were inclined at 45°. A fine mist of water sprayed equally all four samples. Water drops were collected at each slide's base. Water drops were collected at the square array of beads (set 1 unit). With the water drops growing, almost spherical droplets trended to roll downwards when they reached a diameter of ~3.8–4.0 mm. The random array collected an average of 0.95 units and the bare wax sample an average of 0.5 units, with the diameter of droplets about 1 mm; when blown off the edges of the slide, many droplets were lost. Bare glass gave the most variable results: a single large (over 7 mm) flat droplet grew on the surface and ran down the slide by a specific route that varied within each experiment. If the route led to the collecting vial, as in four of the experiments, 1 unit of water was collected. If it did not, as in six experiments, no water was collected.[1]

4.2.2 Hydrophilic Bumps on Polyelectrolyte Multilayers

In some areas such as the Namib Desert, nature has developed excellent methods for water collection from the atmosphere. As shown above, the *Stenocara* beetle in the Namib Desert can collect drinking water from foggy air because of the hydrophilic/superhydrophobic patterned surface of its wings. In a foggy dawn, the *Stenocara* beetle inclines its body forward into the wind and captures small water drops in the fog. When these small water drops coalesce into bigger drops, they slip into the beetle's mouth. Using this method, the beetle is provided a fresh morning drink. Parker et al. found that the structure of the beetle's back resulted in this unique water collecting ability. The *Stenocara* beetle's back is composed of an array of surface bumps on the top with hydrophilic spots on a superhydrophobic background. The small water drops in the morning air head up on the hydrophilic spots and then coalesce. With the water drops growing, the weight of the drops is enough to surmount the binding forces of the hydrophilic region. The water drops separate and run from the superhydrophobic surface to the beetle's mouth. On one side is the amazing water capturing application; and on the other side, the surfaces have applications in the rapid evaluation of complex bioactivities[4–5] and microfluidic channels,[6] which are patterned with hydrophobic and hydrophilic wetting properties. Patterned surfaces with different wetting properties have been used in some techniques, for example, microcontact printing,[7] photolithography,[8] and chemical vapor deposition. However, these methods often involve complicated procedures to add functional groups to the patterned areas or are limited by the range of functional groups. Here, we introduce a simple and more effective route to generate arrays of patterns with chemical functionalities and different wetting properties. Additionally, surfaces with extreme wetting properties, such as a superhydrophobic surface patterned with superhydrophilic designs, can be fabricated by the novel devices, for example, planar microcanals.[9] Open-air microfluidic channels offer advantages such as the facile handling of small amount of liquids, the possibility of massive parallel processing, direct accessibility, and ease of cleaning.[10–11] Application of the patterned surfaces with superhydrophilic and superhydrophobic regions can greatly improve the availability and function of such devices. These also impel us beyond nature's impressive achievement with the Namib beetle. Currently, there is only one report about the patterned superhydrophobic/superhydrophilic surface with limited capability for surface functionalization. A superhydrophobic surface is defined as that on which an advancing water droplet reaches contact angle (CA) of 150° or higher and the CA hysteresis is very low.[12–13] In nature, many surfaces, such as those of the lotus leaf and duck wings, have superhydrophobic characteristics which are result of a multilength scale surface coated with hydrophobic materials.[14–15] Scientists are inspired by these structures to fabricate superhydrophobic surfaces by using a variety of methods.[16–21] Many approaches have been reported to generate superhydrophilic surfaces for applications in self-cleaning.[22–23] Now, we introduce a simple method to generate hydrophilic or superhydrophilic patterns on an otherwise superhydrophobic surface. This method needs to selectively deliver polyelectrolytes to the surface in a mixed water/2-propanol solvent. By using this approach we can fabricate patterned surfaces to mimic the water collection structure of the *Stenocara* beetle's back. Besides, we can introduce specific functional groups or layer-by-layer assembled polyelectrolyte multilayers onto the patterned areas. In this superhydrophobic coatings, rough microporous poly(allylamine hydrochloride) (PAH)/poly(acrylic acid) (PAA) microstructures with PAH/silica nanoparticles (PAH/SiO_2) are covered with a hydrophobic network of semifluorosilane molecules (designated as PAA/PAH/silica nanoparticle/semifluorosilane). If polyelectrolytes could be delivered through the semifluorosilane network, then they would form electrostatic bonds with the underlying PAH or the silica nanoparticles. At the same time, some charged polymer chains would retain on the superhydrophobic surface, which can change the wetting properties of the surface. Using poly(fluorescein isothiocyanate allylamine hydrochloride) (FITC-PAH) forms a hydrophilic domain on a superhydrophobic surface to test this strategy. The superhydrophobic surface is not wettable to a 1% FITC-PAH aqueous solution. But Mohammadi et al. studied that aqueous solutions of various surfactants could help to wet a superhydrophobic surface. This is because the surfactant reduces the surface tension of the solution.[24] Soeno et al. also reported the wetting ability of their superhydrophobic surfaces with a water/ethanol mixture.[25] In this method, water dissolves in FITC-PAH. In the mixture, water (the surface tension of water is 72.2 mJ/m^2) exists a lower surface tension than 2-propanol (the surface tension is 21.7 mJ/m^2).

Micropipeting a pattern of microdrops of 1% FITC-PAH solution onto the superhydrophobic surface forms a hydrophilic region. Then the solvent evaporates. After that, we rinse the patterned surface with deionized water to remove any loosely bound polyelectrolytes. Using the fluorescence intensity of the fluorescein isothiocyanate analyses the density of FITC-PAH on the surface. The result indicates that, after rinsing, the fluorescence intensity of the patterned regions decreases. It means, removal of loosely bound polyelectrolyte. However, due to some FITC-PAH staying on the superhydrophobic surface, the CA and CA hysteresis of the patterned regions change to form a hydrophobic difference between the patterned area and its surrounding field. The relative atomic composition of the hydrophilic domains and superhydrophobic surface can be examined by using the X-ray photoelectron spectroscopy (XPS). The atomic concentration change of nitrogen and the carbon in the patterned area indicate that FITC-PAH molecules on the superhydrophobic surface result in the change of wetting characteristics of the patterned regions. Then, a structure similar to the *Stenocara* beetle's back can be fabricated by patterning an array of hydrophilic area onto a superhydrophobic surface by using a PAA water/2-propanol solution. The hydrophilic patterns seem the wax-free areas on the *Stenocara* beetle's back. It plays an important role in collecting small water drops from the foggy air. Figure 4.2 illustrates the behavior of small water drops on a PAA patterned superhydrophobic surface. The water drops cannot wet the superhydrophobic surface, which leads the small water drops to form nearly perfect spheres. Most water drops roll into the superhydrophobic regions and eventually run into the patterned hydrophilic regions to form larger water drops. The process mimics the *Stenocara* beetle's back to capture the water drops from foggy air. This method can be used to collect small water drops and convert them into larger ones.

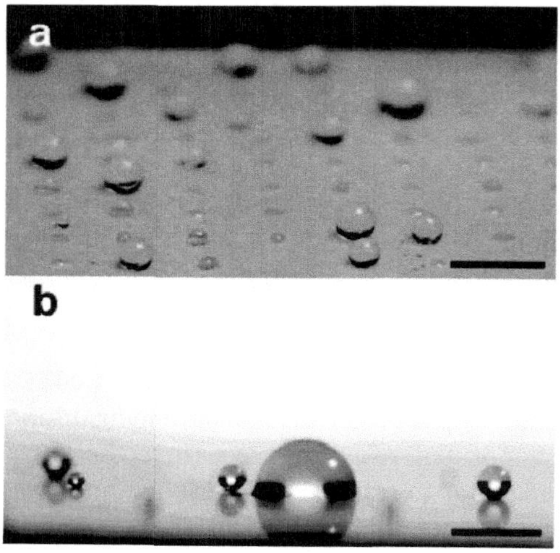

Figure 4.2 (a) Small water droplets sprayed on a (PAA/PAH/silica nanoparticle/semifluorosilane) superhydrophobic surface with an array of hydrophilic domains patterned with a 1% PAA water/2-propanol solution (scale bar = 5 mm). (b) Sprayed small water droplets accumulate on the patterned hydrophilic area shown in (a) (scale bar = 750 μm).

Polyelectrolytes are added to investigate the ability of charged small molecules. These small molecules are bound to a superhydrophobic surface and change the surface properties. The superhydrophobic surface can be patterned with 2-propanol solution of methylene blue (a positively charged dye) and 2-propanol rose bengal (a negatively charged dye) solution with the concentration both of 0.1%. When the dye molecules are placed on the surface, the patterned regions become hydrophobic. Even though the dyes are washed away with water, the areas are still keeping their superhydrophobic properties. This is different from the area patterned with polyelectrolytes. This resulted from the fact that charged small molecules do not have the long polymer chains needed to bind electrostatically with the underlying polyelectrolytes or silica nanoparticles. The 2-propanol conveys some dyes into the semifluorosilane network. Then the dyes are trapped inside the polyelectrolyte multilayer structure. The remaining color in the patterned regions indicates this. The surface can't redissolve the trapped dye using water flushing, because the water is prevented from penetrating into the structure. But if the surface is dealt with hydrophilic feature by using a PAA water/2-propanol solution, these dyes are easily immersed into water from the hydrophilic regions. We can use this method to set chemicals through the superhydrophobic surface into the underlying multilayer structure. The porous polyelectrolyte multilayer structure is like a reservoir for the chemicals. The chemicals can be released to an aqueous medium after creating hydrophilic portals, which has probability to apply in drug release control. An aqueous solution of a chemical acts on the patterned superhydrophobic surface. Then the hydrophobic, different from the superhydrophobic surface and hydrophilic regions, directs molecules to the hydrophilic domains. By using the layer-by-layer technique, the selective assembling of poly-electrolyte multilayers onto the hydrophilic patterns can test this phenomenon. In this process, the nonwetting superhydrophobic background plays a role in resisting the multilayer buildup. Incorporating 0.01% FITC-PAH into the PAH solution monitors the construction of PAH/PAA multilayers. The selective deposition of multilayers onto the patterned surface is confirmed by the large fluorescence intensity contrast between the patterned regions and the superhydrophobic background.

The fabrication method of patterned surface with wetting properties is performed by building superhydrophilic canals on a superhydrophobic background. Superhydrophilic canals, superhydrophilic mutlilayers are created by depositing selectively onto hydrophilic stripes previously patterned on a superhydrophobic surface. The layer-by-layer assembly of PAH and silica nanoparticles coating method has been reported.[23] However, this work uses layer-by-layer deposition scheme reported previously to build superhydrophilic coatings on the patterned area. Figure 4.3 shows the water wetting behavior of microcanals with various numbers of PAH/SiO_2 bilayers by a video CA instrument. A microcanal with four bilayers of PAH/SiO_2 takes more than 10 s to make the water drops spread 1 cm along the 750 μm wide microcanal. This observation indicates that the water cannot be distributed to make the canal superhydrophilic by the capillary force of the coating (Fig. 4.3a). Figure 4.3b shows the microcanal with 14 bilayers of PAH/SiO_2. The water spreading completely along a 6 cm long microcanal only takes less than 2 s. This observation is consistent with the previous study that a stable superhydrophilic behavior needs sufficient number of PAH/SiO_2 bilayers. Because of adding water to a super-hydrophilic microcanal, the water level increasing results in a continuously increasing CA at the edge of the canal. When the CA is more than 90°, a bulge will be formed in the channel and stability is lost. On the contrary, the bulge will disappear when the CA is less than 90°. Gau et al. have reported this observation. Gau and co-workers also studied the construction of liquid bridges between adjacent canals through the coalescence of bulges. These liquid bridges were used to mix the solutions in the canals.[9] However, the liquid bridge can wet the domain between two canals and contaminate the surface without a superhydrophobic background. In this method, due to the canals built on a superhydrophobic surface, the bridge can be broken by simply withdrawing solution from the canals without leaving any solution outside of the canals.

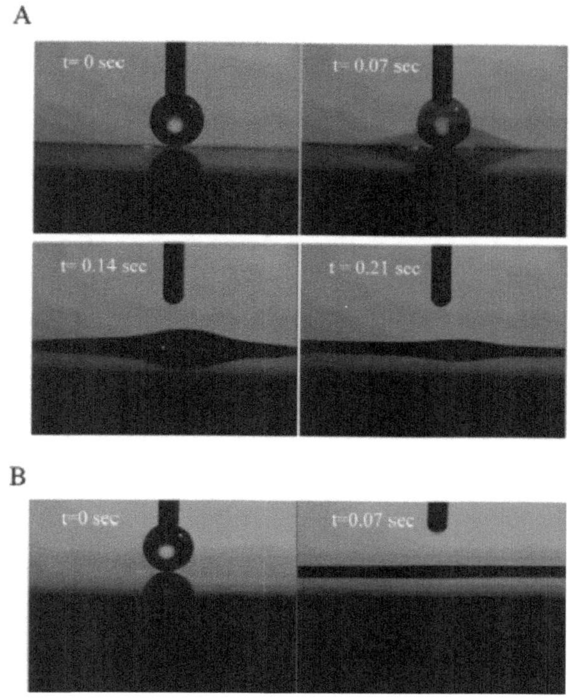

Figure 4.3 The 750 μm wide canals built on a patterned superhydrophobic surface: (A) A water droplet (1.5 mm diameter) spreads along a hydrophilic canal comprised of four bilayers of PAH/SiO_2. (B) A water droplet (1.5 mm diameter) spreads along a superhydrophilic canal comprised of 14 bilayers of PAH/SiO_2.

Patterns with extreme wetting properties can form densely packed small reaction sites for the rapid evaluation of complex biomolecular interactions. In these applications, the patterned region should have a consistent and uniform morphology, and it also should not have interactions between adjacent sites. The high wetting contrast between the patterned area and the background, as well as the cytophobicity and protein adsorption resistance of a superhydrophobic background, makes our patterned superhydrophobic surfaces good candidates for such applications. Figure 4.4 shows an array of circular hydrophilic spots with diameter of 750 ± 30 μm on a superhydrophobic surface. There are various UV-excitable fluorescent dyes. Blue represents 7-amino-4-

methylcou-marin-3-acetic acid, the green is fluorescein, and the red is rhodamine. The dyes are set onto the individual spots. Using this technique, a specific functionalizing reagent could be transferred to an area to explore cell viability, adhesion, and response to different reagents.

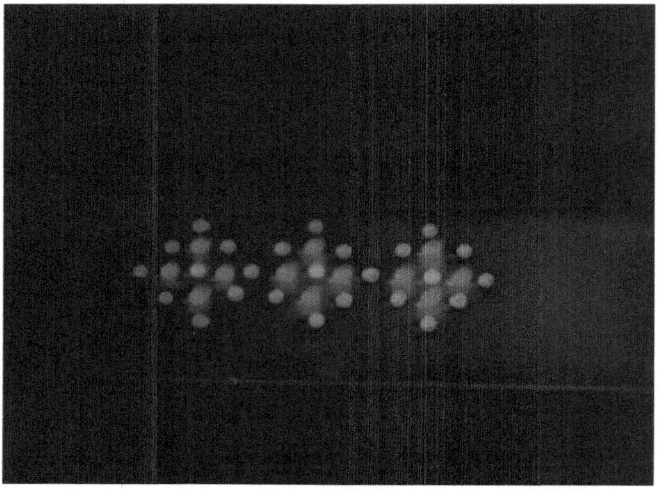

Figure 4.4 Patterned superhydrophobic surface with hydrophilic domains spotted with various dye-containing solutions

In conclusion, the method creating hydrophilic patterns on superhydrophobic surfaces can realize the water collection, which is similar to the Namib Desert beetle. Using this method can introduce some properties and chemical functionalities to the patterned regions by delivering chemicals or multilayer films onto the region. This provides a method to create patterns with desired properties including superhydrophilicity. Many ways can fabricate these patterns to apply in current technologies including inkjet printing, micropipeting, and microcontact printing.[26]

4.2.3 Hydrophilic Bumps on Plasma Polymers

Stenocara beetle lives in arid desert conditions. In desert, the only available source of water is fog droplets. We have shown that the *Stenocara* beetle's back has a random array of smooth hydrophilic bumps. The waxy areas comprising physical features surround the hydrophilic regions, which are arranged in a hexagonal array.

Water collects on the nonwaxy hydrophilic regions. The beetle leans its back wings into the fog to collect the water. When the fog vapor tough the waxy hydrophobic regions, it will blow along the surface until it reaches the nonwaxy hydrophilic region. The water drops grow to cover the entire hydrophilic bump. Due to their own weight, the water drops will roll downward into the beetle's mouth. This method has been mimicked in the laboratory by using random and ordered arrays of 0.6 mm glass spheres on a waxy background. The research shows that an ordered array of spheres can optimize microcondensation. But the CA of only glass is limited to 20° and the CA of waxy hydrophobic background is limited to 111°. Obviously, the degree of hydrophilicity/hydrophobicity and their dimensions play an important role in the overall microcondensation efficiency.

In this section, we will introduce a method to fabricate hydrophilic–hydrophobic patterned substrates, about their surface functionality and dimensional parameters, as well as evaluate the microcondensation efficiency. This method uses plasma-fluorinated polybutadiene and plasma etched poly(tetrafluoroethylene) to employ two types of superhydrophobic background. As Fig. 4.5 shows, a range of hydrophilic plasma polymers have been patterned onto these superhydrophobic backgrounds, and their water microcondensation performances have been compared.

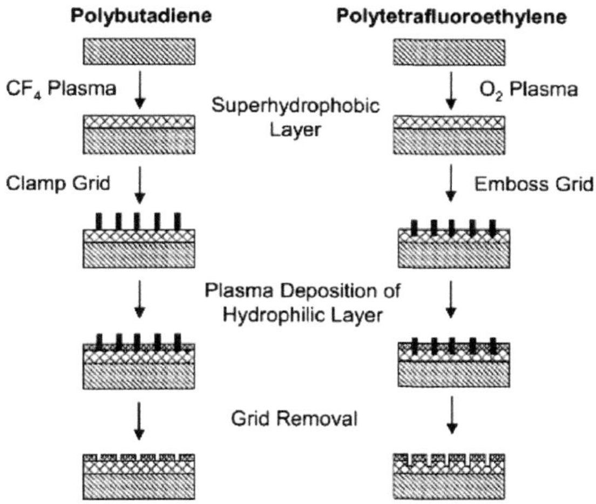

Figure 4.5 Micropatterning a hydrophilic plasma polymer onto a superhydrophobic background (created by either CF_4 plasma fluorination of polybutadiene film, or oxygen plasma treatment of PTFE polymer film).

Badyal et al. reported that microcondensor surfaces can be fabricated by the plasma-chemical arraying of hydrophilic pixels onto a superhydrophobic background.[27] The plasmachemical modification and deposition are carried out in a cylindrical glass reactor. This glass reactor is connected to a two-stage rotary pump by a liquid nitrogen cold trap at a base pressure of 4×10^{-3} mbar and a leak rate better than 5×10^{-3} mol/s. The glass reactor is cleaned using a 50 W air plasma for 30 min. Then the samples are placed into the center of the reactor, which is followed by evacuation to base pressure. Then feed gas/monomer is admitted into the system by a control needle valve at a pressure of 0.2 mbar for 5 min. The electrical discharge is ignited for a selected duration. Then it starts to prepare two types of superhydrophobic surface. An 8% (w/v) solution of polybutadiene in toluene is spin-coated onto polished silicon wafers using a photoresist spinner operating at 2000 rpm for 60 s. These polymer films are vacuum annealed at 90°C for 1 h to remove remaining solvent. Plasma treatment with CF_4 gas at 50 W for 5 min can get a superhydrophobic surface. The plasma-fluorinated polybutadiene-coated silicon wafers are clamped screw-tight to a grid mounted in an aluminum holder, as Fig. 4.6(a) shows. There is another approach to create a superhydrophobic background. It consists of ultrasonic cleaning of 25 mm × 25 mm plaques of poly(tetrafluoroethylene) for 10 min in propan-2-ol and cyclohexane (50/50 v/v) followed by O_2 gas plasma treatment at 50 W for 60 min. These superhydrophobic substrates are embossed with a grid at a pressure of 400 MPa for 10 s using a mechanical press and then mounted in the aluminum holder (shown in Fig. 4.6). The aluminum holder is then placed into the center of plasma chamber in both types of superhydrophobic substrate, which is ready for hydrophilic functionalization through the exposed array holes. The monomer vapor is introduced into the system by a fine needle control valve at a pressure of 0.2 mbar for 5 min before electrical discharge ignition and then evacuation to base pressure. A video capture apparatus measures the sessile drop angle at 20°C. Using a spectrophotometer measures the film thickness. The obtained transmittance-reflectance curves (350–1000 nm wavelength range) are fitted to a Cauchy model for dielectric materials using a modified Levenburg–Marquardt method. Water condenses on the functionalized substrates with high purity water by a nebulizer set to a gas flow rate of 11 L/min N_2. Figure 4.6b shows the process. Then the mass of collected water can be measured.

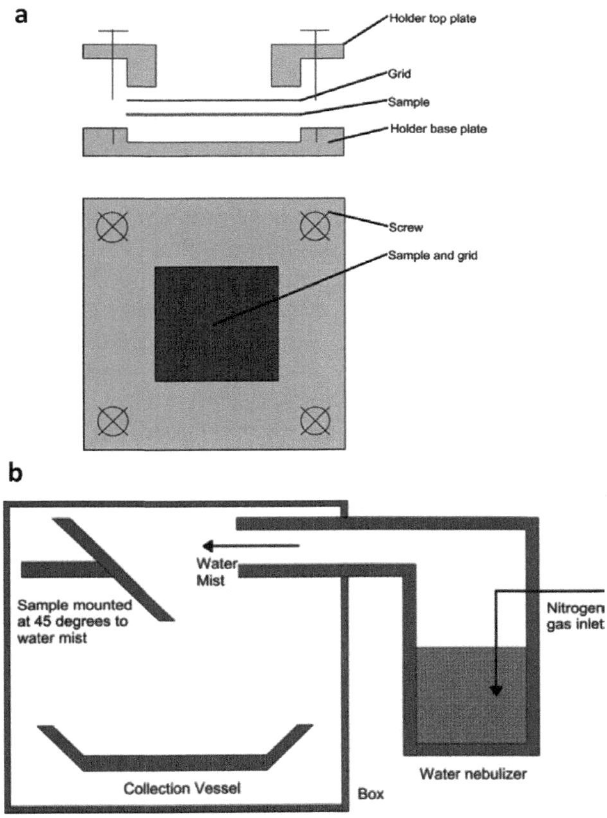

Figure 4.6 (a) Side and top views of the holder used to clamp the grid in position on top of the sample. (b) Apparatus for measurement of water microcondensation efficiency.

This research illustrates that only hydrophilic or superhydrophobic surfaces cannot collect significant volumes of water. Larger water drops are around the edge of the nonpatterned hydrophilic surfaces, and the superhydrophobic surfaces are better to collect water. The cooperation of hydrophilic–hydrophobic surfaces is the most efficient materials for water microcondensation. In this surface fabricated by above method, small water drops condense on the hydrophobic areas. When they grow to reach a hydrophilic region, they will be blown across the surface. With growing on the hydrophilic region, the drops will fall at a steady rate into the collection vessel because of the gravity of a critical size drops.

4.2.4 CVD and Photocatalytic Decomposition

Inspired by the water-collecting strategies used by both desert beetles and spider silks, a novel surface with star-shaped wettability patterns is designed to control motion of tiny droplets that collected on the surface. Particularly, star-shaped wettability patterns integrate both chemical and shape gradients that act on tiny water droplets, realizing highly efficient water collection by introducing a new mechanism. The results provide insights into design of novel materials that can be applied into the realms of water collecting system, wetting control in microfluidics, and so on.

An integrative wettability-controlled surface is designed elaborately by fabricating star-shape angle-regulated wettable gradient following the procedures in Fig. 4.7. At first, a superhydrophilic surface needs to be fabricated by depositing TiO_2 slurry onto a bare glass substrate via a spin-coating method (Fig. 4.7a). And then the superhydrophobic surface is treated with heptadecafluorodecyl-trimethoxysilane (FAS) to realize the change of the wetting property from superhydrophilic (Fig. 4.7a) to superhydrophobic one (Fig. 4.7b). The photomasks with circle-pattern or 4-, 5-, 6-, and 8-pointed star-pattern are used to obtain the shape wettable features via selective illumination of UV light, because after the exposure of UV light, the FAS-modified superhydrophobic TiO_2 surface would be superhydrophilic again due to the photocatalytic decomposition of the FAS monolayer. Thus, the various wettable surfaces with shape gradients are obtained successfully herein (Fig. 4.7c). Although the superhydrophilic surface can spread the droplet (inset of Fig. 4.7a) and the superhydrophobic surface can form the spherical droplet (inset of Fig. 4.7b), the wettable pattern surface can control the behavior of droplets effectively (inset of Fig. 4.7c). Scanning electron microscopy (SEM) is used to observe the microstructure of surface. There is highly porous feature including clusters of TiO_2 nanoparticles of ~50 nm in size and ~19.2 μm in the thickness, respectively. Observation on wettability shows that the unmodified TiO_2 surface is superhydrophilic one with a water CA near to zero, whereas the FAS-modified surface is superhydrophobic one with a CA above 150°. But CAs on the FAS-modified surface change from >150° to <5° after illumination of a UV light (365 nm, ~25 mW/cm^2) for ~60 min. It indicates that the integrative bio-inspired surfaces with wettability patterns (e.g., circle-pattern or 4-, 5-, 6-, and 8-pointed star-pattern) can be formed effectively to control water collection.

It is found that star-shape gradient surfaces realized a more efficient water collection than that of other surfaces such as uniformly superhydrophobic, uniformly superhydrophilic, and circle-shape wettable ones. Especially, the optimal points of star shapes on wettability surfaces would further enhance the efficiency of water collection. Our investigation provides insight into design of novel wettability-gradient surfaces, which may contribute to develop special wettable materials for high-efficiency water or liquid-collecting engineering.

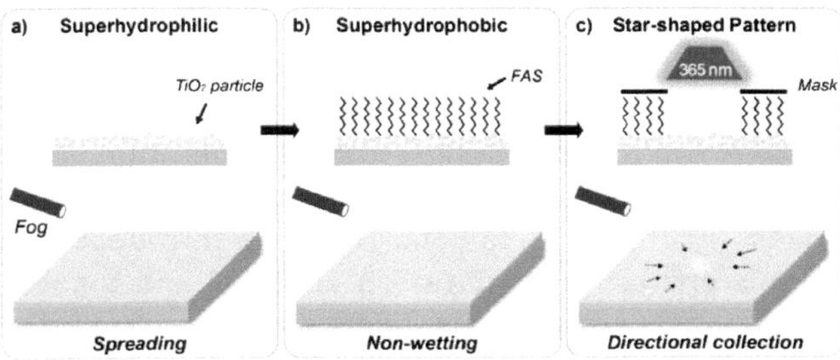

Figure 4.7 Schematic illustration of the fabrication process of bio-inspired surfaces with star-shaped wettability patterns. (a) Superhydrophilic surface composed of TiO_2 nanoparticles, where fog droplets spread (the bottom). (b) Superhydrophobic surface modified by FAS with nonwetting property to fog droplets (the bottom). (c) Bio-inspired gradient surface with star-shaped wettability pattern. It is realized by illuminating the FAS-modified film under UV light with a photomask. The fog droplets would be collected directionally toward the star-shape region, which is more wettable (the bottom).

4.2.5 Hydrophilic Bumps on Polymer Films

It is an increasingly popular method of harvesting water directly from the atmosphere to get the drinking water. In South and Central America, FogQuest utilized a large mesh-like material sheet to collect water from the foggy air. This method has got successful experience in many areas. In this chapter, we will demonstrate the synthesis of micropatterned surface coatings. When the foggy air cools below the surface dew point, the water in the absence of a fog or mist will be captured. Mimicking the mechanism getting from the nature, micropatterns are fabricated with both topographical and chemical features. To enhance the rate of water condensation relative to flat films under representative climatic conditions, the coating are prepared by the dewetting of thin polymer bilayer films on solid substrates[28] to fabricate a surface composed with hydrophilic bumps on a hydrophobic materials. Using this method on a large scale integrated with the existing building infrastructure makes for meeting the water demands of a growing global population, as well as facilitates access to a stable supply of drinking water in remote areas.

There are numerous methods to mimic the water capture mechanism of the *Stenocara*, such as the synthesis of "perfectly hydrophobic" nanofibers,[29] patterning a perfluorinated surface with UV radiation,[30] selective deposition of polyelectrolytes onto a hydrophobic surface,[31] as well as plasma-chemical treatment.[32] All these methods can promote atmospheric condensation. However, it is counteracted by complicated synthetic procedures and the need for specialist synthesis equipment. Here, we introduce a method to create micropatterned surfaces on chemical and topographical regions. Compared with using the spontaneous dewetting of bilayers of thin polymer films, the materials fabricated by this method collect more water. It significantly improves the water drops detachment. Any shape substrates can be used to prepare such coating by conventional techniques, such as dip-coating, spin-coating, and spraying. So it is created with low cost and scalable procedure. It also allows for fabrication on the meter scale.

Unfavorable intermolecular forces at the interface between the two materials drive the unstable thin (~100 nm) liquid films spontaneously breaking apart on a substrate to realize the dewetting process.[33–35] Figure 4.8a shows the formation of micropatterned surfaces via the dewetting of polymer bilayer films. The unstable film eventually forms a series of isolated droplets by breaking apart into holes with time growing. An unstable thin polymer film on a solid substrate spontaneously goes through morphological transformation by dewetting above its glass transition temperature T_g. Below T_g of the polymer film, the droplet morphology pattern is "frozen in." The resultant pattern is used to create biomimetic surfaces and the dewetted polymer droplets are used to mimic the hydrophilic bumps on the *Stenocara*'s back. To create a final surface with wettability contrast between the droplet phase and the background, we show the dewetting of an immiscible polymer bilayer, which is consisting of a hydrophilic polymer layer on top of a hydrophobic polymer under layer.

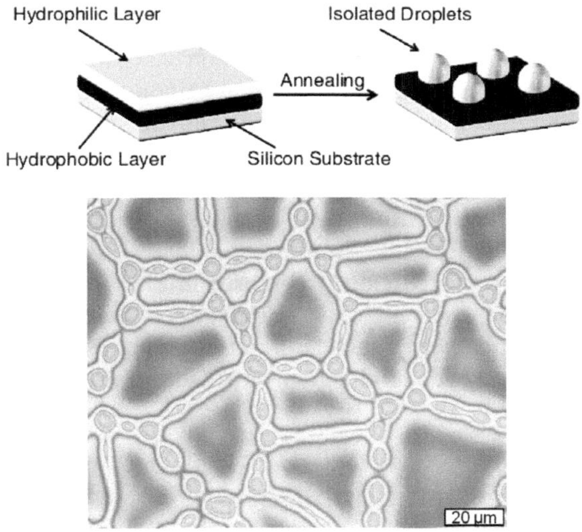

Figure 4.8 (a) Proposed formation of micropatterned surfaces via the dewetting of polymer bilayer films. The top layer is an unstable poly(4-vinylpyridine) (P4VP) film, which spontaneously dewets on the bottom PS layer. (b) Optical microscopy image of a dewetted P4VP-PS bilayer consisting of droplets and interconnected cylinders of P4VP on a PS background.

The spreading coefficient S describes the ability of a liquid to wet a solid surface: $S = \gamma_{\text{solid}} - (\gamma_{\text{liquid}} + \gamma_{\text{solid/liquid}})$, where γ_{solid} is the surface tension of the solid, γ_{liquid} is the surface tension of liquid and $\gamma_{\text{solid/liquid}}$ is the interfacial tension of the solid/liquid interface. We can choose the materials based on their surface and interfacial tensions. If $S < 0$, it shows that the liquid will quickly dewet from the solid substrate. In this method, the polymer bilayer films consist of a top layer of P4VP and a bottom layer of polystyrene (PS). Because of above T_g, S is less than 0. So the P4VP layer can dewet. PS and P4VP has different hydrophobicities (the average equilibrium water CA on flat PS and P4VP films are, respectively, 91.3 ± 1.8° and 36.0 ± 2.7°). The hydrophilicity of P4VP results from hydrogen bonding between water and the lone electron pair on the nitrogen atom in the P4VP monomer unit.

Polymer bilayers are prepared on a clean and smooth silicon substrate. It is consisting of a P4VP top layer and a PS under layer by sequential spin coating. Both two layers are smooth. Annealing temperature is above 160°C (higher than T_g of PS and P4VP).

Figure 4.8b shows the SEM image of the resultant surface morphology that consists of isolated droplets and interconnected cylinders of P4VP on a PS background. When the melt viscosity of the P4VP is lower than that of PS, the P4VP film can be completely dewetting.[36] Preannealing the bilayer can control the height and density of droplet. The temperature should be below the T_g of both layers before dewetting. The preannealing can relax residual stresses in the film and reduce the number of nucleation sites for dewetting. If there are many hydrophilic "bumps" on a hydrophobic background, the surfaces are microscale polymeric mimics of the *Stenocara* exoskeleton. The surfaces are composed by many thousands of hydrophilic bumps per square millimeter. The density is more than *Stenocara*. But the hydrophilic/hydrophobic surface area ratio is still the same.[41]

4.3 Promises and Applications

Hydrophilic–hydrophobic patterned surfaces offer a means of controlling the wetting behavior of aqueous media. Besides this amazing water capturing application, surfaces patterned with different wetting properties (hydrophobic vs. hydrophilic) have applications in microfluidic channels and the rapid evaluation of complex bioactivities. Integrative biosurface with superhydrophilic/superhydrophobic patterns promises much important applications in water-trapping tent, building coverings, and delivering of drugs. Potential applications of such surfaces include controlled drug release coatings[37], open-air microchannel devices, and lab-on-chip devices. This is important for a whole host of technological applications, including cell growth,[38–39] protein manipulation, the spotting of biomolecules,[40–41] microfluidics (to control the location and movement of liquids),[42] and the formation of anti-dew/frost-free protective exteriors. As a possible application for such patterned superhydrophobic surfaces, the collection of water in regions of the earth where fog but no running water is available has been proposed. Other options would be the removing of drops from industrial exhaust gases or the guiding of liquids in microfluidic systems. Surfaces that are (micro)structured into regions of different wettabilities have recently received some scientific interest due to potential applications in microfluidics, microfabrication, and microassembly.

References

1. Parker AR, Lawrence CR (2001) Water capture by a desert beetle, *Nature*, **414**, 33–34.
2. Crawford CS (1981) *Biology of Desert Invertebrates*, Springer, Berlin.
3. Roberts CS, Seely MK, Ward D, Mitchell D, Campbell JD (1991) Body temperatures of Namib Desert tenebrionid beetles: Their relationship in laboratory and field, *Physiol Entomol*, **16**(4), 463–475.
4. Orner BP, Derda R, Lewis RL, Thomson JA, Kiessling LL (2004) Arrays for the combinatorial exploration of cell adhesion, *J Am Chem Soc*, **126**(35), 10808–10809.
5. MacBeath D, Schreiber SL (2000) Printing proteins as microarrays for high throughput function determination, *Science*, **289**(5485), 1760–1763.
6. Lam P, Wynne KJ, Wnek GE (2002) Surface-tension-confined microfluidics, *Langmuir*, **18**, 948–951.
7. Lopez GP, Biebuyck HA, Frisbie CD, Whitesides GM (1993) Imaging of features on surfaces by condensation figures, *Science*, **260**(5108), 647–649.
8. Kim HC, Wallraff G, Kreller CR, Angelos S, Lee VY, Volksen W, Miller RD (2004) Photopatterned nanoporous media, *Nano Lett*, **4**, 1169–1174.
9. Gau H, Herminghaus S, Lenz P, Lipowsky R (1999) Liquid morphologies on structured surfaces: From microchannels to microchips, *Science*, **283**(5398), 46–49.
10. Hsu CH, Chen C, Folch A (2004) Microcanals for micropipette access to single cells in microfluidic environment, *Lab Chip*, **4**, 420–424.
11. Seemann R, Brinkmann M, Kramer EJ, Lange FF, Lipowsky R (2005) Wetting morphologies at microstructured surfaces, *PNAS*, **102**(6), 1848–1852.
12. Chen W, Fadeev AY, Hsieh MC, Oner D, Youngblood J, McCarthy TJ (1999) Ultrahydrophobic and ultralyophobic surfaces: Some comments and examples, *Langmuir*, **15**(10), 3395–3399.
13. Oner D, McCarthy TJ (2000) Ultrahydrophobic surfaces: Effects of topography length scales on wettability, *Langmuir*, **16**(20), 7777–7782.
14. Cassie ABD, Baxter S (1944) Wettability of porous surfaces, *Trans Faraday Soc*, **40**, 546–551.
15. Barthlott W, Neinhuis C (1997) Purity of the sacred lotus, or escape from contamination in biological surfaces, *Planta*, **202**(1), 1–8.

16. Lau KKS, Bico J, Teo KBK, Chhowalla M, Amaratunga GAJ, Milne WI, McKinley GH, Gleason KK (2003) Superhydrophobic carbon nanotube forests, *Nano Lett*, **3**(12), 1701–1705.

17. Feng L, Song Y, Zhai J, Liu B, Xu J, Jiang L, Zhu D (2003) Creation of a superhydrophobic surface from an amphiphilic polymer, *Angew Chem Int Ed*, **42**(7), 800–802.

18. Feng L, Li S, Li H, Zhai J, Song Y, Jiang L, Zhu D (2008) Super-hydropobicity of large-area honeycomb-like aligned carbon nanotubes, *Angew Chem Int Ed*, **41**(7), 1221–1223.

19. Li H, Wang X, Song Y, Liu Y, Li Q, Jiang L, Zhu D (2001) Super-"amphiphobic" aligned carbon nanotube films, *Angew Chem Int Ed*, **40**(9), 1743–1746.

20. Thieme M, Frenzel R, Schmidt S, Simon F, Hennig A, Worch H, Lunkwitz K, Scharnweber D (2001) Generation of ultrahydrophobic properties of aluminium-a first step to self-cleaning transparently coated metal surfaces, *Adv Eng Mater*, **3**(9), 691–695.

21. Erbil HY, Demirel AL, Avci Y, Mert O (2003) Transformation of a simple plastic into a superdrophobic surfaces, *Science*, **299**(5611), 1377–1380.

22. Gu ZZ, Fujishima A, Sato O (2004) Biomimetic titanium dioxide film with structural color and extremely stable hydrophilicity, *Phys Lett*, **85**(21), 5067–5069.

23. Cebeci FC, Wu Z, Zhai L, Cohen RE, Rubner MF (2006) Nanoporosity-driven superhydrophilicity: A means to create multifunctional antifogging coatings, *Langmuir*, **22**(6), 2856–2862.

24. Mohammadi R, Wassink J, Amirfazli A (2004) Effect of surfactants on wetting of super-hydrophobic surfaces, *Langmuir*, **20**(22), 9657–9662.

25. Soeno T, Inokuchi K, Shiratori S (2004) Ultra-water-repellent surface: Fabrication of complicated structure of SiO_2 nanoparticles by electrostatic self-assembled films, *Appl Surf Sci*, **237**, 543–547.

26. L Zhai, Michael CB, Fevzi C, Yushan K, John MM, Michael FR, Robert EC (2006) Patterned superhydrophobic surfaces: Toward a synthetic mimic of the Namib Desert beetle, *Nano Lett*, **6**(6), 1213–1217.

27. Garrod RP, Harris LG, Schofield WCE, McGettrick J, Ward LJ, Teare DOH, Badyal JPS (2007) Mimicking a *Stenocara* beetle back for micro condensation using plasma chemical patterned superhydrophobic-superhydrophilic surfaces, *Langmuir*, **23**(2), 689–693.

28. Reiter G, Hamieh M, Damman P, Sclavons S, Gabriele S, Vilmin T, Raphael E (2005) Residual stresses in thin polymer films cause rupture and dominate early stages of dewetting, *Nat Mater*, **4**, 754–758.

29. Chen R, Zhang X, Su Z, Gong R, Ge X, Zhang H, Wang C (2009) Perfectly hydrophobic silicone nanofiber coating: Preparation from methyltrialkoxysilanes and use as water-collecting substrate, *J Phys Chem C*, **113**, 8350–8356 .

30. Dorrer C, Ruehe J (2008) Mimicking the *Stenocara* beetle-dewetting of drops from a patterned superhydrophobic surface, *Langmuir*, **24**, 6154–6158.

31. Zhai L, Berg MC, Cebeci FC, Kim Y, Milwid JM, Rubner MF, Cohen RE (2006) Patterned superhydrophobic surfaces: Toward a synthetic mimic of the Namib Desert beetle, *Nano Lett*, **6**(6), 1213–1217.

32. Garrod RP, Harris LG, Schofield WCE, McGettrick J, Ward LJ, Teare DOH, Badyal JPS (2007) Mimicking a *Stenocara* beetle's back for microcondensation using plasmachemical patterned superhydrophobic-superhydrophilic surfaces, *Langmuir*, **23**, 689–693.

33. Neto C, Jacobs K (2004) Dynamics of hole growth in dewetting polystyrene films, *Physica A*, **339**, 66–71.

34. Seemann R, Herminghaus S, Jacobs K (2001) Gaining control on pattern formation of dewetting liquid films, *J Phys Condens Matter*, **13**, 4925–4938.

35. Seemann R, Herminghaus S, Neto C, Schlagowski S, Podzimek D, Konrad R, Mantz H, Jacobs K (2005) Dynamics and structure formation in thin polymer melt films, *J Phys Condens Matter*, **17**, S267–S290.

36. Thickett SC, Harris A, Neto C (2010) Interplay between dewetting and layer inversion in poly(4-vinylpyridine)/polystyrene bilayers, *Langmuir*, **26**, 15989–15999.

37. Stuart CT, Chiara N, Andrew TH (2011) Biomimetic surface coatings for atmospheric water capture prepared by dewetting of polymer films, *Adv Mater*, **23**, 3718–3722.

38. Ito Y (1999) Surface micropatterning to regulate cell functions, *Biomaterials*, **20**(23–24), 2333–2342.

39. Nygren N (1996) Initial reactions of whole blood with hydrophilic and hydrophobic titanium surfaces, *Colloids Surf B*, **6**, 329–333.

40. Gilmor SD, Thgiel AJ, Strother TC, Smith LM, Lagally MG (2000) Hydrophilic/hydrophobic patterned surfaces as templates for DNA arrays, *Langmuir*, **16**, 7223–7228.

41. Zhang G, Yan X, Hou X, Lu G, Yang B, Wu L, Shen J (2003) Binary DNA arrays on heterogeneous patterned surfaces, *Langmuir*, **19**, 9850–9854.

42. Handique K, Burke DT, Mastrangelo CH, Burns MA (2000) Nanoliter liquid metering in microchannels using hydrophobic patterns, *Anal Chem*, **72**, 4100–4109.

Index

An environmentally friendly book printed and bound in England by www.printondemand-worldwide.com

PEFC Certified

This product is from sustainably managed forests and controlled sources

www.pefc.org

PEFC/16-33-415

This book is made of chain-of-custody materials; FSC materials for the cover and PEFC materials for the text pages.